Urban Government for the Paris Region

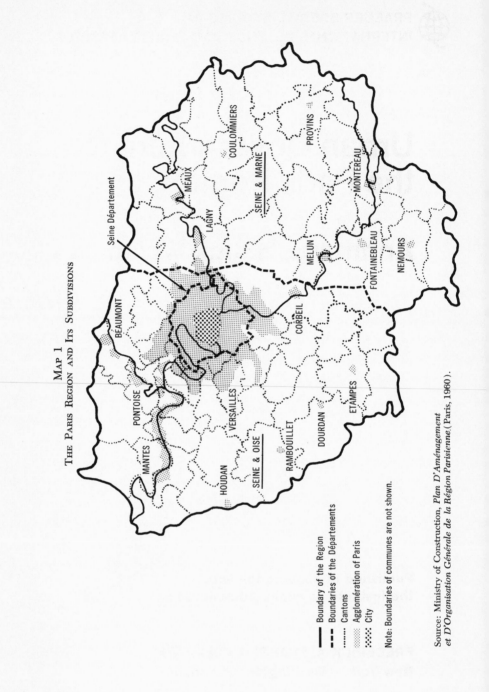

MAP 1
THE PARIS REGION AND ITS SUBDIVISIONS

Seine Département

COULOMMIERS

PROVINS

MEAUX

MONTEREAU

SEINE & MARNE

LAGNY

BEAUMONT

MELUN

FONTAINEBLEAU

NEMOURS

CORBEIL

PONTOISE

VERSAILLES

ETAMPES

MANTES

DOURDAN

HOUDAN

RAMBOUILLET

SEINE & OISE

— Boundary of the Region

｜ Boundaries of the Départements

‥‥ Cantons

Agglomération of Paris

City

Note: Boundaries of communes are not shown.

Source: Ministry of Construction, *Plan D'Aménagement et D'Organisation Générale de la Région Parisienne* (Paris, 1960).

PRAEGER SPECIAL STUDIES IN
INTERNATIONAL POLITICS AND PUBLIC AFFAIRS

Urban Government for the Paris Region

Annmarie Hauck Walsh

Published in cooperation with
the Institute of Public Administration

FREDERICK A. PRAEGER, Publishers
New York · Washington · London

The purpose of the Praeger Special Studies is to make specialized research monographs in U.S. and international economics and politics available to the academic, business, and government communities. For further information, write to the Special Projects Division, Frederick A. Praeger, Publishers, 111 Fourth Avenue, New York, N.Y. 10003.

This book is No. 1 in the series *The International Urban Studies of the Institute of Public Administration.*

FREDERICK A. PRAEGER, PUBLISHERS
111 Fourth Avenue, New York, N.Y. 10003, U.S.A.
77-79 Charlotte Street, London W.1, England

Published in the United States of America in 1968
by Frederick A. Praeger, Inc., Publishers

© 1968 by the Institute of Public Administration

Library of Congress Catalog Card Number: 67-16664

Printed in the United States of America

FOREWORD

This study is one of a series of twelve
similarly structured case studies of urban government carried out by the Institute of Public Administration in 1964-66. The Institute's research program in international urban studies, supported by
the Ford Foundation, is intended to provide raw
material for the young field of comparative urban
administration. By following the same detailed research outline in examining urban areas in Africa,
Asia, Europe, North America, and South America we
have developed comparable data and descriptive
analyses of the structure and issues of urban government and public administration in nations with
widely varying political, economic, and cultural
systems. This approach should help students and
scholars to develop comparative insights and pro-
positions on the administrative aspects of urbanization. As another part of the project, IPA has
prepared for the Division for Public Administration
of the United Nations a comparative review of administrative aspects of urbanization, with emphasis on the interests of less developed nations
(to be published by Frederick A. Praeger in 1968).

This study of Paris is based on examination
of documentation such as government budgets, plans,
annual reports, and laws, and on interviews with
officials in the Paris Region. In identifying
problems and issues of urban government, we have
been guided by locally expressed values and policy
goals. Paris proved to be one of the most excit-
ing cases, for she is in the midst of major governmental adjustments in response to common problems of urban development. While the field work
was conducted during 1964-65, important events of
1966 are noted in this book.

The help of many persons who have contributed
to the study is gratefully acknowledged. Particular thanks go to Yves Sabouret, Inspecteur des
Finances, who undertook a substantial part of the

v

research on Paris and has aided the author in dis-
covering and interpreting the facts presented, and to
Michel Picquard and Peter Wengert who have gener-
ously contributed advice and assistance. Participants in the project at the Institute included
Joseph E. McLean and Randy H. Hamilton (former
project directors), Frank Smallwood, and Rodman
T. Davis. The advice of Emil J. Sady of the United
Nations and suggestions from members of the Comparative Urban Studies Section of the Comparative
Administration Group (American Society for Public
Administration) have been valuable throughout. The
project secretary, Anthony Asalone, has been indis-
pensable to whatever success has been achieved.

Annmarie Hauck Walsh, project director, not only
has prepared this book, but has been responsible
for examination of foreign language materials and
for review of the work in several case study
areas.

Lyle C. Fitch
President
Institute of Public Administration
New York

CONTENTS

LIST OF MAPS, CHARTS, AND TABLES

MAPS

CHARTS

TABLES

agglomération
: a metropolitan area as statistically defined in France by criteria of density of population, continuity of construction, and communications with the central city. The Paris agglomération includes the central, built-up portion of the Paris Region (10 per cent of its area and 90 per cent of its population).

aménagement du territoire
: town and country planning; planning, controlling, and developing land and uses of it.

arrondissement
: a territorial subdivision of a département that is the administrative jurisdiction of a sub-prefect. The average arrondissement includes about 150 communes. The same term applies to the twenty wards of the City of Paris, each of which contains a town hall and an administrative subordinate of the Prefect of Seine, called a mayor.

canton
: a territorial subdivision of a normal arrondissement utilized primarily as an election district for departmental councils.

central city
: the major urban municipality of an agglomération; in the case of Paris, it is the City of Paris.

CGP
: the Commissariat Général du Plan, or General Planning Commission of France, which is situated in the prime minister's office.

commune	the uniform legal municipal unit throughout France, of which the major institutions are elected council and mayor.
<u>DAT</u>	the Délégation à l'Aménagement du Territoire et à l'Action Régionale, or Office for Town and Country Planning and Regional Action, which is situated in the prime minister's office and is charged with helping to develop urban and regional planning policies and overseeing national and regional plan implementation.
Delegate General (Délégué Générale) to the District of the Paris Region	the delegate of the national government to the Paris Region, who is executive of the district itself and adviser to the prime minister on policy for the region. As of 1966, the post was merged into that of regional prefect.
<u>département</u> (departmental)	the intermediate level of administration in France at which both branches of the central ministry services and directly elected councils, with their subordinate services, are found. There are ninety <u>départements</u> in France, of which three--Seine, Seine-et-Oise, and Seine-et-Marne--are wholly situated within the Paris Region. Under a 1964 reorganization bill, these three will be restructured into seven <u>départements</u> by January 1, 1968.

District of the Paris Region	a special district with jurisdiction over the entire Paris Region, which has planning and financial-aid responsibilities. It consists of a governing council, the delegate general, and the delegation or staff services.
field services	operating agencies of national ministries organized at regional, departmental, and local levels throughout France.
government	in French usage refers to the Executive of the nation--President, prime ministers, ministers, and their cabinets.
HLM	Habitations à Loyer Modéré, moderate rental housing units and the local government or cooperative agencies that construct them under terms of national legislation.
IAURP	Institut d'Aménagement et d'Urbanisme de la Région Parisienne, or Paris Region Institute for Planning and Urbanism, which is a public enterprise undertaking research and planning work, much of it on behalf of the District of the Paris Region.
MRP	Mouvement Républicain Populaire, a national political party of the center.
PADOG	Plan d'Aménagement et d'Organisation Générale de la Région Parisienne, the general physical plan prepared by SARP (Paris Region Office for Town and Country Planning) in 1960 and replaced by a new regional plan prepared by the district in 1965.

xiv

Paris Region   the legally defined jurisdiction
                      of the district, which is the
                      greater urban area of Paris, in-
                      cluding three départements and
                      over 1,300 communes.

prefect   the chief executive and chief
                      delegate of the national govern-
                      ment in each département, who is
                      responsible to the Ministry of
                      Interior.

prefecture   the administrative agencies and
                      officers under the authority of
                      the prefect.

RATP   Régie Autonome des Transports
                      Parisiens, the regional transport
                      authority, which provides major
                      transit services in the Paris
                      urban area.

régie   a public corporation directly
                      controlled by one or more
                      government units.

SARP   Service d'Aménagement de la
                      Région Parisienne, or Paris Re-
                      gion Office of Town and Country
                      Planning, which is a field ser-
                      vice of the Ministry of Construc-
                      tion.

secretary-general   a professional chief administra-
                      tive officer responsible to the
                      chief executive, utilized in
                      departmental and communal admin-
                      istrations, the prime minister's
                      office, and the Ministry of
                      Education.

SFIO   Section Française de l'Inter-
                      nationale Ouvrière, a national
                      political party in the socialist
                      group.

SNCF	Société Nationale de Chemin de Fer, the public corporation that provides all railroad services in France.
state (état)	the broad term referring to the government of France represented by both central-government agencies and officials, and officials on lower levels when they are acting as agents of the national government.
syndicat	a special authority created by several governmental units for joint provision of one or more public services or functions. Frequently an intermunicipal special district.
UDT	Union Démocratique du Travail, a national political party allied with the Union pour la Nouvelle République.
UNR	Union pour la Nouvelle République, a national political party supporting President de Gaulle.

Urban Government for the Paris Region

CHAPTER **1** DIMENSIONS OF
THE REGION

The Paris Region dominates France, and Paris
dominates its region. As the unrivaled cultural,
economic, and administrative capital of the nation,
Paris has invoked both pride and jealousy from the
citizens of the traditionally rural nation, reac-
tions that have had direct impact on the condition
and development of governmental services in the
capital.

The City of Paris is endowed with monumental
beauty and the richest supply of public services
in France from hundreds of years of French history
of which she is the star. While the urban complex
has grown steadily and spread rapidly in the postwar years, however, the rate of public investment
has remained far behind the expansion and modification of needs for public plant and services--for
housing, transportation, water supply and sewerage,
and community facilities. The tendencies of national political pressures and related government policies during this period were toward braking the
growth of the capital region and channeling development, both public and private, to other parts of the
nation in order to reduce the dichotomy between
"Paris and the French desert."* For the provinces
were bent on reducing the near monopoly of the Paris
Region in services, cultural facilities, higher education and research, and major industries.

---

*This is the title of a well-known book by J.-F.
Gravier, which in 1947 urged more balanced distribution of population and economic activities in France.

Comprising only about 2 per cent of the nation's land area, the region has almost 20 per cent of its population. More significantly, the region claims roughly 40 per cent of all professional and top-management personnel and over 30 per cent of service, middle-executive, and white-collar employees. Business headquarters, credit institutions, labor leadership, top-level public administrations, and important industries, such as electricity and automobile manufacturing, are concentrated in the Paris Region.

Government structure for the City of Paris and Département of Seine, immediately surrounding it, differs from that common to local government throughout the rest of France. The capital is subjected to far greater control and direct administration by national authorities. Special legislation on municipal and departmental organization, personnel, and various public activities applies to it. The President and Prime Minister of France have participated directly in much major and some minor decision-making for the urban complex. During the past five years, new structures have been instituted throughout France to organize public activities at a regional level. In this respect also, the Paris Region has been subject to a special approach.

Until very recently, the dominant attitude of the national government toward Paris throughout the postwar period was that constraints upon it were the most important inducement to a better national distribution of population and economic activities. This view held that the less the investment in housing, transport, schools, and related facilities in Paris, the greater the tendency of families and enterprises to settle elsewhere. The shortages in these categories are today more severe in the Paris Region than in other urban areas of France.

In fact, the Paris Region has continued to grow since 1945, simultaneously with accelerated urbanization in other parts of the nation. By 1962, almost 40 per cent of the French population resided in

agglomérations (which are comparable to Standard
Metropolitan Statistical Areas in the United States)
of 50,000 or more.* Twenty-two of these areas have
experienced growth rates higher than that of Paris;
for example, between 1954 and 1962 growth was 44.5
per cent in Grenoble, 35.3 per cent in Besançon,
35.2 per cent in Caen, 26.3 per cent in Dunkerque,
compared with 15 per cent in the Paris agglomération.

Recent residential growth of the Paris area has
been concentrated in the suburbs, which include small
communes with minimal government structure, where the
gaps between public services and demand for them have
become greatest. The concept of an urban region as a
unified market and policy jurisdiction has been slow
to develop; rather, the central city itself has been
the focus of public and private investment through the
years. The conurbation is composed primarily of the
central city, on the one hand--with its high quality
and selection of public and private facilities and
rich urban life--and suburban towns, on the other
hand--with little more than a post office, a bank
branch, a secondary school, and some small-scale
factory and commercial employment. Within the region, centralization of employment and urban leisure
facilities, together with over-all scale of growth,
has produced major urban problems requiring region-
wide action, particularly in the realm of transportation and development planning.

By 1960, the deficiencies in public facilities
and the cumbersomeness of the administrative machinery in the region were sufficiently obvious to spot-
light two facts: First, the Paris urban area

---

*The Paris agglomération (7.4 million people) is
the only agglomération of over 1 million in France.
There are 3 from 500,000 to 999,999 (Lyons,Marseille,
and Lille-Roubaix); 6 from 300,000 to 499,999; 26 from
100,000 to 299,999; and about 185 from 20,000 to 99,999.
The degree of urbanization in France is low for Western
Europe.

must be planned and equipped for fast-growing population with fast-increasing wealth, simultaneously
with development of the rest of the nation; and
second, this approach requires modification in concepts of the scale and interrelationships of urban
problems and of appropriate government structures to
deal with them.

National executive stability established under
the Fifth Republic created a favorable environment
for organized efforts to consider systematically the
problems of the Paris Region and alternative solutions
to them. The years 1960-65 have produced a flurry of
activity on this front, including governmental reorganization, planning and research, and stepped-up
public investment in the urban area.

Because the activities of national authorities
are so preponderant a part of administration for the
Paris Region, the major thrust of reorganization
measures for Paris has been the development of
institutions that facilitate formulation of national
policy and coordination of national activities at the
metropolitan or regional level. These include creation of the District of the Paris Region, with comprehensive planning and investment functions; appointment of a regional prefect; and utilization of the
Interministerial Committee for the Paris Region.
In addition, modifications in local administration
have been made, including some delegation of powers
to departmental prefects and municipal councils and
the restructuring of the three départements in the
region into seven. Economic and land-use plans have
been prepared by new institutions and form the basis
for major investments, particularly in transportation
and water-supply facilities, and satellite cities.
The period since 1958 is thus an era of significant
change and innovation in the Paris urban complex
that will influence urban form and governmental procedures for many decades to come. A remarkable
aspect of these changes in comparison to similar
efforts in other two- or multi-party nations is the
degree to which they have been effected with little
overt public, party, or intergovernmental conflict.

President de Gaulle, in announcing the establishment of the regional prefect, stated: "It is necessary to centralize the direction of administration
[for the Paris Region] and to institute responsibility within it." Others have disagreed with this
imperative but have not vigorously opposed measures
taken in its name.

The central concern of this study is to identify
the official actors engaged in planning and governing the Paris urban area, and to examine the structure of governmental institutions involved, their
formal roles, and their interrelationships. The
structure includes institutions at the three traditional levels of French government: National government, or the State; the département, of which three
have been wholly included in the Paris Region; and
the municipality, or commune, of which there are
1,305 (including the City of Paris) in the region
(see Map 1). In addition, new regional structures
form a fourth tier of activity.

The remainder of Chapter 1 deals briefly with
the demographic, economic, and political environment
in which these institutions operate. In Chapter 2,
the institutions are identified and described.
Subsequent chapters examine intergovernmental relationships, the administration of planning and plan
implementation, and organization for providing four
selected urban services--water supply, mass transportation, public housing, and education. Finally,
a comparative summary of administrative structures
and problems that relates the Paris urban area to
the urban areas studied in other volumes of this
series is presented in Chapter 6.

THE POPULATION AND THE ECONOMY

The Paris Region is an area of almost 4,700
square miles and 8.5 million inhabitants. Nearly
90 per cent of this population, however, resides
in 10 per cent of the area--the urbanized 465

square miles, or agglomération, in the center of
the region (see Map 1).* Encompassed by the region
are the three départements of Seine, Seine-et-Oise,
and Seine-et-Marne, which include the City of Paris
and 1,304 other municipal units, or communes. The
agglomération covers 221 communes, and over 900
communes in the region are rural units with populations under 2,000.

The population of the City of Paris has remained fairly stable in postwar years, but that of the
region has increased by over 16 per cent in the last
decade (national growth was about 12 per cent). The
annual growth rate of the Paris Region is presently
about 1.85 per cent.

The population of the city grew steadily until
1921, after which the growth of the inner suburbs of
Seine, the central département surrounding Paris, was
intense until World War II. In the last ten years,
rapid growth has been concentrated in the suburban
portions of the agglomération--the outer portions of
Seine, the inner segment of Seine-et-Oise, a small
part of Seine-et-Marne--and further in strips along
the valleys of the Seine, Marne, and Oise rivers and
around major highway and railroad axes. Many of
these segments have experienced growth of 50 to 150
per cent over a twenty-five year period. Planning

---

*For the purposes of this study, the "urban
area" refers to the Paris Region as statutorily defined, and described below. The agglomération is
the central built-up portion of the region, defined
by the Institut National des Statistiques et des
Etudes Economiques by criteria of density and continuity of construction.

The population of the region, according to
estimates for 1964, was distributed as follows:

Paris agglomération	7,400,000
Other urban centers in region	300,000
Peri-urban towns	300,000
Rural areas	500,000
	8,500,000

TABLE 1

POPULATION INCREASE IN FRANCE
(in thousands)

Year	Nation	Paris Region	Département of Seine	City of Paris
1946	40,282	6,598	4,766	2,700
1954	42,777	7,317	5,175	2,850
1960	45,335	8,197	5,573	3,035
1965 (est.)	48,000	8,500	6,000	2,850

authorities for the urban area now consider a projection of 14 million people in the region by 1990
to 2010 to be conservative.

The outermost parts of the region have lost
population to the agglomération, and residential
concentration remains high; in 1962 the density of
persons per square mile was 1,853 in the region as
a whole, 15,500 in the agglomération, and 83,000 in
the City of Paris. This central density is considerably higher than that of other world urban areas
of comparable size.

Migration into the Paris Region accounted for
about 60 per cent of its growth from 1954 to 1962.
The majority of the migrants are young people who
come directly from other urban places to seek improved employment opportunity or career advancement.
The predominant pattern of French migration is movement from rural birthplace to a local urban center
to Paris. Slightly over half of the migrants from
the provinces settle in the city, but many later
move to the suburbs. Paris assimilates migrants
swiftly; their occupational and locational profiles
resemble those of native Parisiens.[1]

The age distribution of the urban area is
similar to that of the nation (about 55 per cent
under forty), and there are about 113 females per
100 males (as compared with 108 females to 100
males in the nation). Average household size is
lower in the urban area, with about 75 per cent of
the region's households consisting of one, two, or
three persons. The region's birth rate is 17.1 (as
compared with 18.3 for the nation), and its death
rate is 9.7 (as compared with 11.2 for the nation).
Literacy rates are close to 100 per cent. The population is ethnically homogeneous, with the exception
of a sizable minority of citizens of North African
origin (60,000 in Seine), naturalized citizens
(170,000 in Seine), and foreigners (213,000 in Seine)

The economy of the region is far more diversi-

fied than that of other French urban areas. The
industrial (secondary) and commerce-services (tertiary) sectors are relatively balanced, although
the latter is expanding at a faster rate and both
are internally diverse. In 1962, the resident labor
force was employed as follows:*

	Labor Force	Per Cent
In primary activities	64,000	1.6
In secondary activities	1,727,000	43.7
In tertiary activities	2,167,000	54.7
Total	3,958,000	100.0

Slightly over 20 per cent of the labor force of the
region is employed in the public sector, including
general government and public enterprise.

Employment in the Paris Region as a whole is
fairly concentrated. While industry is scattered
in the inner suburbs and radially along major transportation axes, commerce and services are intensively centralized in the city, with some scattering
of business headquarters to the west. Total employment is more centralized than residential settlement, resulting in an average daily commutation toward the center of the region by over 2 million persons.

───────────────

*"Primary" includes agriculture, forestry, and
mining; "secondary" includes manufacture and assembly;
"tertiary" includes commerce (wholesale and retail
marketing) and services.

Analysis of the municipalities of the urbanized portion of the region by a public enterprise
engaged in urban research* has confirmed what strikes
the observer's eye: Surrounding the City of Paris
are concentric rings (albeit imperfect) of varying
types of development. The innermost ring is old,
highly industrialized, and dense and contains relatively low-income population. The second ring has
highly mixed characteristics: somewhat higher standard of living, median industrialization, and a
mixture of high-rise construction and single-family
homes. The third ring is rapidly growing and has a
young population with relatively low standard of
living and intense commutation toward the center.
The fourth ring resembles the third but has a lower
rate of commutation to the center. In contrast to
most large metropolitan areas in the United States,
the central city in Paris encompasses relatively
high average incomes and low blue-collar labor
force, and its rate of depopulation to the suburbs
is slow.** Expansion of automobile ownership and
suburbanization of middle-income groups are only
recent phenomena in Paris.

The governmental plans for the Paris Region
that are percolating at present contemplate modify-
ing the spatial structure of the economy by establishing self-sufficient urban centers in the outer
rings, on the one hand, and further concentrating
business headquarters in a modern central business
district, on the other.

---

*Institut d'Aménagement et d'Urbanisme de la
Région Parisienne, "Comparaison et classification
des communes de l'agglomération parisienne," Cahiers
(Paris, 1965), Vol. 3. This study was based on factor
analysis of some sixty-nine indicators, which were
synthesized into six factors: balance of employment
and population; urbanization and industrialization;
commutation toward Paris; standard of living; demographic structure; and growth rate.

**These characteristics are shared, however, by
most of the other urban areas abroad that were
examined in this research effort.

TABLE 2

WORKING POPULATION BY PLACE OF WORK IN THE PARIS REGION,
BY SELECTED ACTIVITIES, 1954 CENSUS
(in thousands)

	Paris	Seine Suburbs	Seine-et-Oise	Seine-et-Marne	Total
Construction and Public Works	74	71	64	16	225
Other Industries	498	504	263	53	1,318
Transport	76	66	46	9	197
Commerce, Banking, Insurance	373	215	126	26	740
Services	268	114	81	18	481
Public Administration (Inc. Army)	188	131	88	20	427
TOTAL	1,477 (44%)	1,101 (33%)	668 (19%)	142 (4%)	3,388 (100%)

Unemployment in the region is extremely low by international standards.* The number of unfulfilled requests for jobs registered in the region with labor ministry offices ranged from 20,790 to 30,832 in 1964. Moreover, wages and salaries in Paris are higher than those in the rest of France (where the average monthly earnings of employees are 777.5 francs and of laborers are 717.8 francs).** The standard of living has been increasing rapidly (by approximately 50 per cent in the past decade in France as a whole) and consumption levels have risen steadily.*** Paris Region inhabitants spend 18.7 per cent of their consumption outlay on transport, vacations, and cultural and leisure activities, while other Frenchmen so dispose of 14.5 per cent.

The general picture of the region, then, is one of a growing, literate, and employed urban population with modestly high average incomes on an international scale, young in age and newly suburbanizing, and faced with expanding economic opportunities.

---

*Occupational distribution in the region was found in the 1954 census to be as follows: agricultural workers, 89,000; patrons (directors, proprietors), 416,000; professions and top management, 217,000; middle management, 370,000; employees (salaried), 728,000; laborers, 1,423,000; personal service, 324,000.

**In adjusted dollars, these figures approximate $157 and $145 respectively. The source of them is INSEE, Etudes et conjoncture, supplément no. 5 de 1965.

***See INSEE, Annuaire statistique abrégé de la région parisienne (Paris, 1961). The wage index for laborers in the region (based on 100 for 1949) was 307 at the end of 1960, while the price index (also based on 100 for 1949) in the central city was 195.

POLITICS

Paris politics are a microcosm of French politics (which is not true of local politics in the rural and small-town sectors of the region and of the rest of France, where local issues and local loyalties are vigorous). The political game in the city and the Département of Seine has engaged all major shifting parties and party groups of the Fourth and Fifth Republics. Elections to municipal and département councils are the testing grounds for national elections, and these offices are often steppingstones to parliamentary office.

Throughout France there are, underlying the fluid and complex party structures, fairly stable attitude cleavages--pro-Church and anticlerical, peasant and urban, right and left, Communist and anti-Communist--which are imperfectly reflected in the party structures. The parties function less as dominant interest-aggregating mechanisms than as generally undisciplined, loosely organized channels for recruitment of the elected officials. Party membership is low (about half a million out of the 19 million voters in France were registered party members in 1962). Most political groups are distinguishable more by leading personalities and fine philosophical differences than by stands on real and present policy issues.

The major groupings in the National Assembly elected in 1962 include the Republican Independents, the Gaullist parties, the democratic center, the Radical Socialists (who are neither radical nor socialist), the socialists, and the Communist Party. The Republican Independents (with thirty-three deputies) are a loosely bound group on the right with a base in moderate-income groups with nostalgia for monarchial and peasant structures. Some of its members support the incumbent government. The major Gaullist party is the Union pour la Nouvelle République (UNR), which has been bolstered since 1962 by the Union Démocratique du Travail (UDT), a small liberal subgroup supporting the government.

The UNR was forged from members of the center and
right (particularly from the postwar party of
de Gaulle, the Rassemblement du Peuple Français) on
the basis of support for General de Gaulle, who is
its undisputed leader. The UNR-UDT combination
claims 229 of the 465 deputies elected in 1962 and
controls the present government. The core of the
democratic center is the Mouvement Républicain Popu-
laire (MRP) which has strong support from adherents
of Catholicism and has high representation in the
bureaucracy. It supports the separation of church
and state, is progressive in social and economic
philosophy, and has stronger national organization
than most French parties. It has, however, little
influence in the Paris Region. The Radical Socialist Party (thirty-eight deputies) also of the
center, is a federation of local committees ("meetings of notables") and was the major party in the
Third and Fourth Republics. The socialist group
(sixty-four deputies) includes the Section Française
de l'Internationale Ouvrière (SFIO) and various splin-
ter groups that have support from wage earners and
white-collar workers in the north and urban areas
of France, but it has not expounded strong programs for social and economic change. The Communist
Party is an exception among French parties in its
tight organization and high membership (300,000).
Over half its membership is in the Paris Region.
While its expressed domestic program is vague and
its electoral power has declined with the successes
of de Gaulle, its presence strongly affects the
pattern of political conflict.

Temporary alliances among parties for local
and département council elections are common. In
the Paris Region, the SFIO is allied sometimes with
the Communist Party, sometimes with the MRP. The
MRP is allied sometimes with the SFIO, sometimes
with the UNR. In the local elections of March,
1965, the socialists and the Communists allied
against the UNR in many local districts in the
Paris area.

Political issues in local elections in the

Paris Region are predominantly national, not only
because of the importance of these elections in the
national political arena, but also because of the
centralization of Paris government and the nationally oriented perspectives of Paris citizens. Urban
policy issues are more frequently settled in national ministries than in local councils. The debates
of the 1965 elections in Paris were cast in terms of
Gaullism and anti-Gaullism. The current administrative reorganization of the region did not become an
important election issue. Voting abstentions were
high--about 35 per cent of registered voters in the
city did not vote.

There is some variance in party dominance among
parts of the region, generally following social and
economic lines. The UNR is strong in the central
city. The Seine suburbs are a Communist stronghold.
The Radical Socialists have some strength in the
rural communes of the region. In the 1962 elections
to Parliament, roughly 43 per cent of votes in the
City of Paris were cast for the UNR. The Communist
Party received 38 per cent of the votes in the Seine
suburbs. The Paris City Council after the 1965
local elections, however, included 38 Gaullists, 39
members from the Communist-socialist coalition, and
13 members from center groups, with none having a
working majority.

The stands taken by the various parties on
issues of local government and reorganization are
closely related to their bases of power. The
Gaullists, who are generally a minority power on the
municipal level, have supported administrative reforms that tend to heighten centralization of government for Paris. They have been far from monolithic,
however. In addition, the government-sponsored reorganization of the Paris Region into seven départements and the separated City serves to break up the
Communist-socialist strength in Seine, although it
creates a smaller Communist unit. The Communist
Party has often adopted popular stands on behalf of
"local liberties," as has the SFIO. The Radical
Socialists generally defend the status quo in the

rural communes of their strength. In France generally, defense of local powers against central
authority, however, transcends party lines and is
centered in the Association of Mayors (which, para-
doxically, is itself a highly centralized organization) and in the group of mayor-senators in the
upper house of the national legislature.

The French political system is in a state of
flux. The stable majority in the first decade of the
Fifth Republic, recent electoral reforms eliminating
proportional representation in municipalities of over
30,000 population, and the institution of direct
elections for the national presidency generate pressures toward polarization and simplification of the
party system. While the opposition parties have not
consolidated, the socialists and radicals joined a
federation of the left for the 1967 legislative
elections and entered into an alliance with the
Communist Party for the second ballot. As a result,
the Gaullist parties fell just short of a majority
although they polled nearly the same proportion of
popular votes as in 1962. In any case, the parties
have not served to clarify urban policy issues in the
Paris Region or to crystallize opinion on them. This
has minimized political restraints on administrative
decision-making in urban affairs.

The formation and activity of nonparty interest
groups--including labor groups, professional associations, religious and educational groups, and
peasant associations--have increased in the postwar
period. Such groups, however, exert relatively
little direct pressure on local government. Every
important national group has a Paris office and
focuses its attention on the national administration
and legislators. Specialized government agencies
are frequently client-oriented and respondent to
such pressure. Religious and labor groups also work
through several political parties. There are parliamentary rules against deputies' supporting particular
interest groups, but local interests frequently request legislators to intercede with administrative
bureaus on their behalf. Major organizations rely
also on methods of direct communication aimed at
the government, both written and active. (Strikes and

demonstrations are common tactics of farmers, public
service employees, and students, for example.)

In the few discernible instances in recent
years of specific interest-group pressures opposing
urban policies of the administration in Paris, the
groups have had little success. Business and financial groups vehemently opposed the moving of the
major produce and wine markets from the central city,
to no avail. In the case of the wine market, the
move was a personal decision of Prime Minister Debré.
On the other hand, business and labor interests are
represented on government advisory committees ranging in importance from the Economic and Social
Council, which works on the national four-year
plan, to specialized committees for technical
agencies.

Although civil servants cannot hold elective
or party office while active, they can do so on
temporary leave from the service. Conflicts of
interests between administrative and political roles
are clearly recognized in French administrative
practice. On partisan political issues, administrators have an obligation of neutrality in dealing
with the public and carrying out policies of the
government. This principle extends to the private
lives of top-level administrators as a general
obligation of restraint. These obligations have
strong roots in tradition and administrative training and are enforceable in administrative law. On
the other hand, a broad range of practical decision-
making takes place within the bureaucracy in France.
Neutrality refers to party competition, group favor-
itism, and broad political goals, not to practical
policy formulation. Regional urban policies for
Paris have been formulated in large part within the
bureaucracy.

The balance between political responsibility
and administrative independence is a delicate one
that is nevertheless sustained to the extent that,
with the exception of ministerial cabinets, there
is negligible change in administrative personnel at

all levels of government when party control changes.
Political control of the central bureaucracy is
exercised primarily through the ministers and prefects. In general, the bureaucratic establishment
has far greater powers relative to political officials, particularly on the local levels, in the
French system than is found in Anglo-Saxon or
Eastern European systems. A mayor cannot dismiss
municipal personnel, for they are protected by
national regulations that afford them job security,
even though they are recruited and paid by the
commune. At all levels of government, personnel
discipline is handled through the administrative
courts. While politically responsible mayors and
local councils direct local staff as to what to do,
how it is to be done often either is set forth in
national regulations or is decided within the
administrative structure.

Requirements of loyalty to the political
leaders entail execution of policy and orders given,
but the senior civil-service staff, which in the
postwar period has been stable during the rapid
turnover of governments, exercises fully its rights
to persuade and influence the political leaders.
Staff alliances with political interests do occur,
and the relationship between a minister and division head, or a mayor and his secretary-general,
vary considerably with the personalities of the
participants.

In summary, the major participants in decision-
making on urban issues in Paris are political officials and bureaucratic groups. In the Paris Region,
the general population, political parties, and
interest groups have been noticeably passive in the
face of major changes in government structure and
public investment. Yet, in a recent public-opinion
poll in the region, three out of four people said
that they disliked living in a large urban complex.
They held little hope for major improvements in the
public sector of urban life, but focused ambition
on improvements in private wealth. A new planning
organization--the District of the Paris Region--is

setting out to awaken public consciousness to
regional problems and to stimulate formulation of
goals for improvement. Such an awakening, if it
occurs, could stir the whole structure of politics
in the region. The party structure, expansion of
private wealth, centralization of power, and popular attitudes have thus far, however, lulled the
sleeping giant.

Notes to Chapter 1

1. See Guy Pourcher, *Le Peuplement de Paris*
(Paris: Presses Universitaires de France, 1964).

Government for the Paris Region involves institutions at four jurisdictional levels--national,
regional, departmental, and municipal--in an essentially unitary but complex system (see Chart 1).
The national government, or State,* not only directly
undertakes public-service activities in the region
through its ministries and field agents at the
regional, departmental, and local levels, but also
supervises the local units of government. While the
major governmental institutions belong to the four
distinguishable tiers and are described in such
categories, the structural concept of separate levels
of government does not apply to the behavior of the
total government system.

Two historical streams of policy have shaped
the present local-government system in France:
that tending to a centralized, hierarchical administration directed by centrally appointed executives,
based on elements inherited from the monarchy and
systematized by Napoleon; and that stressing the
rights and powers of elected councils, emanating
from the revolution and developed throughout the
nineteenth century. There are cultural factors
reinforcing both tendencies; prevalent belief in
the necessity for strong and focused authority in
human affairs--in family, religion, and society--
coexists with a fundamental distrust of authority
and with a strong spirit of local independence in
France. While the centralization of government in
France is often exaggerated--for local government

---

*Etat is the broader term in French usage repre-
senting the permanent government of the nation, including lower-level officials when they are acting as
national agents. For purposes of clarity, the term
"national government" is utilized more frequently in
this study and is distinguished from departmental
and municipal government on the basis of the controlling budget (that voted by Parliament, a departmental
council, or a municipal council).

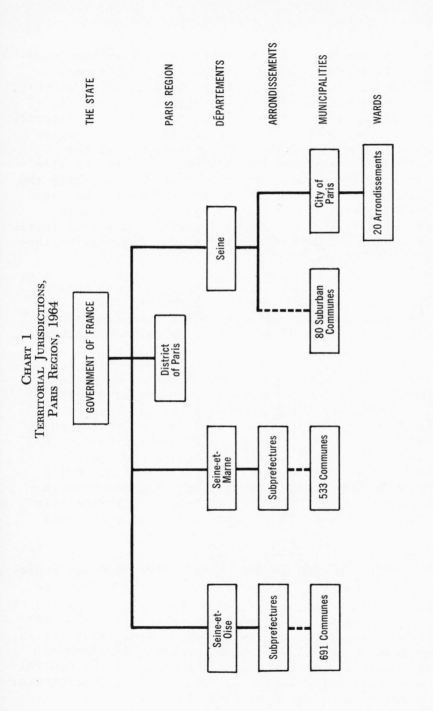

CHART 1
TERRITORIAL JURISDICTIONS,
PARIS REGION, 1964

THE STATE

PARIS REGION

DÉPARTEMENTS

ARRONDISSEMENTS

MUNICIPALITIES

WARDS

GOVERNMENT OF FRANCE

District of Paris

Seine-et-Oise

Seine-et-Marne

Seine

Subprefectures

Subprefectures

City of Paris

691 Communes

533 Communes

80 Suburban Communes

20 Arrondissements

21

is vital throughout the country--there is no question that it prevails in Paris due to the City's
special administrative system, arising from its
national importance and the political antipathy of
the provinces to its power.

The Constitution of the Fifth Republic says
only the following on local government:

> The territorial units of the Republic
> are the communes, the Départements, the
> Overseas Territories. Other territorial
> units may be created by law.

> These units shall be free to govern
> themselves through elected councils and
> under conditions stipulated by law.

> In the Départements and the Territor-
> ies, the Delegate of the Government shall
> be responsible for the national interests,
> for administrative supervision, and for
> seeing that the laws are respected.

Two principles expressed in this brief statement underlie the framework of local government.
First, commune and département councils are organs
of a national system, and while they have general
rights to govern their own constituencies, their
powers and functions depend upon the conditions
laid down by national law and decrees as enforced
by supervising executives. There have been no
major proponents in French history of local "autonomy" or home rule. Decentralization measures have
expanded the powers of local authorities over a
century but have never presumed to give them exclusive decision-making prerogatives over local matters.
Second, the executive--and territorial agents of
the executive--have powers independent of legislative bodies. Police power (broadly construed as
responsibility for the protection of public safety,
health, and morals) and management of public services
are executive functions, constitutionally and statutorily assigned to appointed ministers and prefects

and elected mayors. The national executive is law-
maker on matters not defined by organic law as loi
(basic law, which is the province of the national
legislature). Even on these matters, the legislature may delegate powers to the executive, which
may then issue decrees.

Historically a centralized system, administration in France has undergone two major types of
modification. Through delegation, certain governmental powers and functions have been shifted to
field offices and agents of the central government.
Later to begin, devolution of powers to local authorities has paralleled the growth of representative
institutions at the communal and département levels,
particularly early in the Third Republic, when
elected mayors were instituted and specific functions were delegated to them. The end product is a
pattern of overlapping bureaucratic and representative structures. The national field offices are
placed primarily at regional and département levels.
Representative organs are the departmental and municipal councils. Their corresponding chief execu-
tives--prefects and mayors respectively--act both
as executives of the councils and as agents of the
national administration, depending on the activity.

Below the national level, therefore, there are
three types of regular administrative jurisdiction
in France. The first type is the pure field division of national services, in which all authority
and major policy flow from the superior hierarchical level. There are several of these in the Paris
Region, such as the regional division (academie) of
the Ministry of Education and a regional branch of
a Ministry of Construction planning service. These
are, for the most part, single-function units with
coterminous or overlapping areal jurisdictions. The
second type of jurisdiction encompasses both field-
division and local-government units with distinc-
tive powers but shared executive and/or administrative institutions. Belonging to this category are
the départements of which the prefect is both executive of the departmental council and coordinator of

national field services. Budgetary authority flows
both horizontally from the council and vertically
from national ministries to the administrative
structure at the département level. The third type
is the jurisdiction of an independent local-government unit, such as a commune or municipality, in
which budgetary authority flows horizontally for the
most part.

NATIONAL INSTITUTIONS AND
THE GOVERNMENT OF PARIS

The role of Parliament in governing Paris has
traditionally been a large but general one. Legislation determines the structure of government for
the region, the general powers of local institutions,
and the requirements for approvals by agencies of
the national administration of local-government
activities. Because Paris is treated apart from
other regions of France in special legislation, the
parliamentary role has greater direct impact on the
capital. The general outlines of current reorganization of the Paris Region have been embodied in
parliamentary bills. The details, however, are
filled in by executive decrees. More important,
general policies regarding the development of Paris
in relation to the rest of the nation, while formulated by the executive, reflect the pressures arti-
culated in the legislature. The effort to channel
investment to other sections of the nation, for
example, has been a response to the demands of the
province-dominated Assembly. Specific legislative
constraints on administration policies for Paris
arise from time to time when interregional issues
are involved. Thus, for example, opposition in the
Assembly to use of Loire Valley water sources for
the City of Paris underlies the resistance of the
government to this proposal by city water authorities.

Particularly under the Fifth Republic, however, dominant power on most major issues concerning Paris rests with the government (executive)
and administration.

The effective power of the executive has expanded in practice, supported by the dominance of
the UNR, but the constitution of the Fifth Republic
does restrict parliamentary control and strengthen
executive regulatory powers relative to that of the
Fourth. The executive fixes the agenda of the legislature, which must produce a motion of censure signed
by one tenth the membership and supported by majority
vote to bring down the government. Any bill or
resolution designated by the prime minister as a
matter of confidence is considered passed if a
motion of censure against it is not passed, which
gives government bills a strong advantage.

The Government

Officially the government consists of the
prime minister (head of the government), some fifteen ministers and additional ministers of state,
together with ministerial secretariats. The prime
minister is responsible for management of the administration, execution of laws, national defense,
and appointments (except those specifically assigned to the President). The Council of Ministers and
the prime minister are collectively responsible for
government policy. The President of the Republic
does not have constitutional powers to run the
administration under normal conditions. He presides over the Council of Ministers, however, signs
executive decrees and orders, and appoints the
prime minister and top civil and military officers.
In earlier republics, the President seldom inter-
vened in the decisions of the government. Under
the Gaullist government and legislative majority,
however, as sustained by the dominant personality
of the incumbent President, it is difficult to distinguish between presidential and governmental
roles in the Fifth Republic. The President set
the wheels in motion for reorganization of Paris
and has personally participated in substantive
policy decisions for the region. Direct election
of the President subsequent to 1962 may extend this
tendency beyond the de Gaulle era, but to date it

has been in large part a function of the incumbent's influence and capacity for leadership.

The immediate prime minister's office includes the government secretariat and the two national-planning units. The ministers, who are appointed by the President on the prime minister's recommendation, head permanent ministries or perform special tasks as ministers of state (without portfolio). A minister of state for administrative reform, for example, has been appointed to participate in a special policy-development effort that led to, among other measures, the 1964 administrative reorganization of the Paris Region. (Primary responsibility for formulating Paris reorganization measures was assigned in 1963, however, to the Minister of Interior, Roger Frey.)

Traditionally, ministers have been political appointees holding parliamentary seats. Under the Fifth Republic, however, parliamentary and government office cannot be held concurrently. About a third of the ministerial posts in the government of Prime Ministers Debré and Pompidou have been held by career civil servants.

In addition to participating in policy-making in the Council of Ministers, each regular minister is directly responsible for the management of his ministry. He has the immediate aid of a cabinet of personal appointees, usually including senior general administrators of the civil service, technicians, political aides, and recruits from private enterprise. This staff provides information and advice, coordinates the work of branches of the ministry, and often has considerable delegated authority to control administration.

During the Third and Fourth Republics, the technical ministries (and particularly their bureaucratic structure, or the "administration"*)

---

*The "administration" in French usage refers

enjoyed considerable independence. Coordination
of ministry activities on national, regional, and
département levels, however, has been an active
concern of the current government. The requirements for developing comprehensive policies and
concerted investment programs for the Paris Region
have highlighted coordination problems of a system
in which the traditional independence of ministries,
and even of particular services and special civil
service corps, is great. Certain of the coordinating mechanisms--the powers of the President and
prime minister, national planning efforts, the hege-
mony of the Ministry of Finance--apply to the nation
as a whole. Others--the Interministerial Committee
for the Paris Region, the regional prefect for Paris,
and regional plans--are explicitly concerned with
policy and coordination on the level of the urban
region. These are all innovations of the Fifth
Republic, since 1958. Coordination of national
government activities within the département is
formally the role of the prefect.

The President and prime minister (particularly
Prime Minister Debré) have had direct impact on
ministerial activities in the Paris area because of
their intervention in decisions for the region.
Relatively technical decisions relating to Paris
that would normally be taken within the administration have been made by the President or prime minister in council or in the Interministerial Committee.

The primacy of the Ministry of Finance and
Economic Affairs has a long history and is firmly
established. It has responsibility for budgetary
and fiscal policy and for control of public expenditure by all public authorities--including other
ministries. In exercising this power, finance authorities tend to go beyond the issues of legality to

---

to the permanent senior officers of the ministries,
their subordinate officers and agencies, as distinguished from the politically responsible
"government." There are no political executives
in French government below the level of ministerial cabinets.

questions of advisability of expenditures. Comptrollers (contrôleurs des dépenses engagées) responsible to the Ministry of Finance but attached to
the various spending ministries pass on all expenditures with respect to budgetary allocation, availability of credits, and conformity to financial
regulations. The minister of finance arbitrates in
case of disputes. Moreover, the budget division
negotiates with the various ministries over their
estimates in drawing up the annual budget and financial law, and treasury officials execute the actual
financial transactions within each ministry. The
finance ministry also manages various investment
funds (such as the Economic and Social Development
Fund) that participate in national loans, subsidies,
and loan guarantees. As is generally true of centralized financial control in other countries, it
has been exercised in France more to assure economy
and fiscal regularity and to apply general fiscal
policy than to harmonize substantive policies and
programs of various ministries and agencies. The
conservative stance of the Ministry of Finance
toward Paris added a brake on public expenditure and
investment in that region during the fifteen years
following World War II.

The Interministerial Committee for the Paris
Region, which was established in the 1960's by
decree, is charged with developing policies and
coordinating the activities of the ministries as
they affect the Paris Region. It includes the
ministers of interior, finance, construction, and
public works and transport; the secretary of state
for the interior; and the prime minister, as chairman. The regional prefect of Paris serves as its
secretary and often its informal leader. He prepares its agenda and presents dossiers on problems
to be discussed. Other ministries participate on
an ad hoc basis. The powers of the committee are
consultative, and it did not have much impact on
the administration in its early years. Because
the government had not formulated comprehensive
policies for the region that could form the basis
for interministerial coordination and because

tendencies to ministerial independence were persistent, the committee's importance developed slowly.
At present it meets regularly and has become the
focus of the most important national decisions on
development of the region. Although not formally
empowered, it is the most significant decision-
making body with respect to investments and major
works in the region. This development is in large
part attributable to the will and power of the President and his determination to utilize the committee
to guide activities of the ministries in terms of
general policies for the region, as well as to the
plans of the District of the Paris Region, which
are urged on the committee by the regional prefect.

The Administration

The administrative agencies of the central
ministries perform some public activities directly
in the Paris Region and supervise other activities
of commune and département authorities. The Ministry of Interior is responsible for general supervision of local government, but the functional
ministries exercise technical supervision over particular projects and services, particularly when
national financial participation is involved.

Directly subordinate to the minister and his
cabinet are general administration divisions (e.g.,
personnel, budget, purchasing) and manifold functional directorates. The precise form of hierarchical organization below this level varies considerably from ministry to ministry, but the common
subunit with clear-cut functional boundaries is the
bureau. In some cases there are services on a line
between the directorates and bureaux; other services
are separate from the directorate organization and
responsible directly to the minister.

The ministries have a wide network of regional
and local branches, or "field services," which are
territorial subdivisions of ministry directorates
(they may be services or bureaux). The territorial units are not uniform in scale, but all major

field services have agencies at the level of the
département in the Paris Region and some are organized at the regional level (see Chart 2). Some of
these services operate quite apart from local authorities (e.g., army, justice, post office), while
others are closely associated with them (e.g., education, bridges and roads, treasury), sometimes
undertaking services financed by local or département councils and frequently providing supervision
and technical assistance. (In rural areas they often
implement the bulk of local authority services.)

Most ministries, in addition to central and
field services and bureaux, have inspectorates,
either general or specialized by branch, the staff
of which examines private institutions (e.g., construction and factory inspectors) or local authorities (e.g., public-finance inspectors). Some of
these function as line officials, such as the education inspectors, who are the direct supervisors of
secondary- and primary-school teachers. Finally,
advisory councils, which include officials of various agencies and representatives of private interests and local authorities, are frequently attached
to ministerial agencies. Consultation with these
councils is often a requisite step in project establishment by local or central services, although
their opinions are only advisory.

The Civil Service

The staffs of the ministries and their field
services, and the prefects and their immediate
staffs, are members of the national civil service.
Personnel of the municipal and département councils
throughout France, outside Paris and Seine, are
separate local cadres. In the postwar period, however, national statutes and decrees on local-government personnel have regulated job classification,
salary scales, rights, and qualifications in order
to upgrade local staff and facilitate mobility by
equalizing conditions.

In 1964, the senior administrative staffs of

the City of Paris and Département of Seine were sub-
sumed under the national civil service. Throughout
the postwar period, the officers in this grade had
waned by attrition in Seine. Before 1962, improvements in pay and conditions made in the national
civil service had not applied to Seine staff, and
that département, like other local units, could not
recruit from the Ecole Nationale d'Administration.
Understaffing had become severe, and intensive negotiations and litigation in the administrative courts
ended in the extension of the national civil service
to Seine, further intensifying the differences in
structure between the capital and the other local
units in France.

In the national civil service, executive and
clerical personnel are recruited by the various
ministries and prefects. Senior administrative
officers, however, are uniformly recruited through
the Ecole Nationale d'Administration, and higher
technical officers through the Ecole Polytechnique.*
Technical civil servants, with engineering and
scientific training, are promoted to top general
management of the technical ministries and agencies
(such as transport, construction, and agriculture).
Specific categories of administrative staff form
special cadres, to which leading applicants are
recruited toward the end of their graduate training
and in which they remain, regardless of assigned
post. These include the grands corps--Inspection
des Finances, Cours de Comptes, Conseil d'Etat; the
senior technical corps--corps of mining engineers

---

*These schools provide a form of in-service
training, as students are recruited into them, paid
by the State while in attendance, and have civil
service obligations upon graduation. The three-year
course at the Ecole Nationale d'Administration includes graduate courses in law, economics, public
administration and finance, as well as supervised
work in a field post.

CHART 2
ADMINISTRATIVE FRAMEWORK, PARIS REGION, 1964

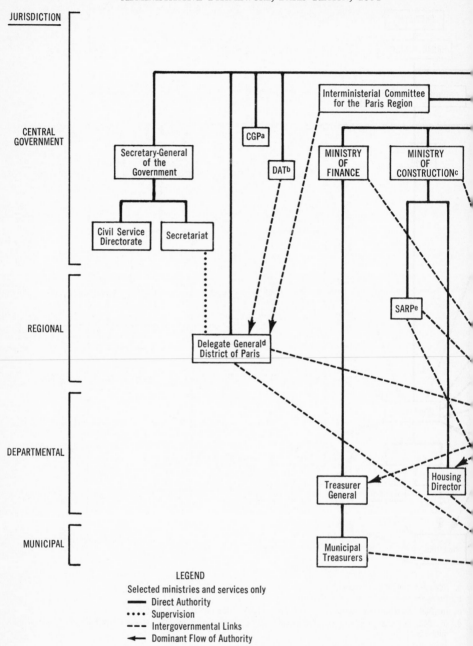

LEGEND

Selected ministries and services only
— Direct Authority
•••• Supervision
--- Intergovernmental Links
◄— Dominant Flow of Authority

Source: Adapted from F. Ridley & J. Blondel, *Public Administration in France* (London: Routledge and Kegan Paul, 1964).

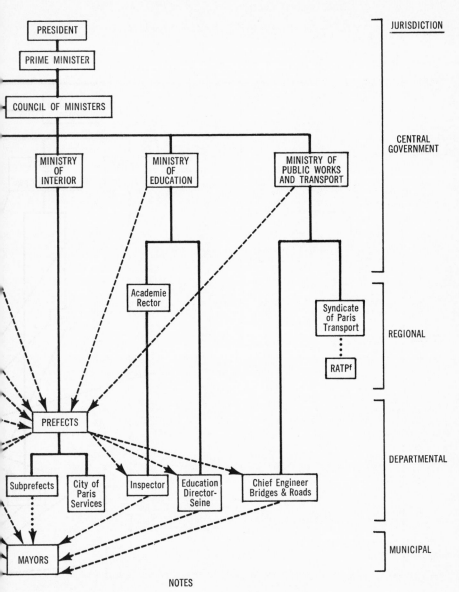

JURISDICTION

CENTRAL GOVERNMENT

REGIONAL

DEPARTMENTAL

MUNICIPAL

PRESIDENT

PRIME MINISTER

COUNCIL OF MINISTERS

MINISTRY OF INTERIOR

MINISTRY OF EDUCATION

MINISTRY OF PUBLIC WORKS AND TRANSPORT

Academie Rector

Syndicate of Paris Transport

RATPf

PREFECTS

Subprefects

City of Paris Services

Inspector

Education Director-Seine

Chief Engineer Bridges & Roads

MAYORS

NOTES

a. General Planning Commission.
b. Office for Town and Country Planning and Regional Action.
c. In 1966, the Ministry of Construction became a branch of the Ministry of Public Works.
d. In 1966, this post was merged with that of regional prefect, who will be responsible to the minister of interior.
e. Town and Country Planning Service for the Paris Region.
f. Paris Region Transport Authority.

33

and corps of civil engineers (bridges and roads); and the prefectoral corps. They have high prestige and spirit of independence, and they fill all types of top government posts. Upon resignation from government service, they also make up a significant proportion of the executive and technical personnel of private enterprise.

The elite status of the higher civil service, its rapport with private enterprise, its reputation for competence and honesty, and its firm tradition of service to the State have in general reinforced the role of the bureaucracy that is formally based on extensive executive powers.

The Council of State

The highest administrative court in France maintains judicial control of administration as legal arbiter over intergovernmental relationships and conflicts between citizens and government agencies. Questions of legality of government activity and proper roles of various government bodies are settled by court law. Further, the council advises the government on the legality of policy proposals and administrative regulations and may, on its own initiative, suggest legislative and administrative reforms. The importance of administrative law and the Council of State has significant ramifications for the whole government system as it operates in Paris and elsewhere. Special authorities, government contracts, new government programs, and allocation of responsibility among agencies of government on both the national level and the local level are regularized through administrative case law. Hence, it is understandable that the training of the administrative class of the civil service has strong legal content.

The ramifications of this system of administrative law for urban administration are significant. Many issues that become the stuff of intergovernmental policy conflict in other areas are resolved by case law in France. In addition, the

legal framework stabilizes the exercise of higher
government control over local authorities by defining the scope of discretionary control.

THE DEPARTEMENTS IN THE PARIS REGION

Up to 1964 there were three départements in the
Paris Region. In July, 1964, Parliament passed a
government-sponsored bill reorganizing the departmental divisions into seven. The changes are being
instituted gradually, and the new system is scheduled to be functioning by the beginning of 1968. The
subject for analysis at this point in time is primarily the pre-existing system.

Of the three traditional départements in the
urban area, Seine-et-Oise and Seine-et-Marne have
been structured and empowered under general administrative law in much the same way as the other départements in France (of which there are ninety).
The Département of Seine, in which the City of Paris
and its inner suburbs are situated, has been subject
to special laws and is an exception to the uniformity of departmental structure, just as the City of
Paris is an exception to the uniformity of municipal
structure. The reason for these exceptions has been
explicitly expounded by central officials of France
throughout history. The dominance and potential
power of Paris--the generator of political as well
as intellectual and technical revolution in France--
are considered too great to escape administrative
fetters. As Haussmann, the renowned nineteenth
century prefect of Seine, phrased it, "The organization of an autonomous Paris under any form whatso-
ever, could only be creation of a State within a
State." All of the organizational reforms of the
Fifth Republic (regional institutions, département
reorganization, changes in personnel systems, and
planning efforts) have confirmed and strengthened
the tradition of special treatment for Paris and
Seine.

The major governmental institutions in each

département are the prefect and prefecture administration and the departmental council (conseil général).

The Role of the Prefect

The prefect of an average département combines
aspects of a political and an administrative official, of a national and local officer. He is
appointed by the government, is subject to instant
dismissal, and is directly responsible to the minister of the interior. He is generally a career civil
servant, however, of known loyalty to the incumbent
government, and while turnover of prefects has taken
place in history with changes of governments, this
is rare.

As agent of the national government, the prefect is the representative of all ministers and is
titular head of technical and administrative services in his département (with the exception of
justice). His approval is necessary for all payments from the national budget in the département.

The field-service branches of central ministries
in each département are theoretically responsible to
the coordinating and supervisory powers of the département prefect, who is the general delegate of
the government in that jurisdiction. The growth,
however, of the separate technical services at the
departmental level, each responsible to the counter-
part central authorities in the ministries, has considerably weakened the coordinating role of the prefect with respect to national administration in the
département. The various service branches in Seine
constituted largely unrelated functional "fiefs."
Government moves in recent years to coordinate
administration for purposes of plan implementation,
both at regional and departmental levels, included
a series of decrees of March, 1964, that transfer
to the prefects decision-making powers formerly
exercised by service chiefs (with certain exceptions in finance, education, and labor administration). Outside the Paris Region, prefectoral approval is required henceforth for major decisions by

heads of departmental branches of national services,
and future delegation of powers from central ministries to their departmental officers must flow
through the prefects. Prefects, in turn, can delegate the decision powers on specified matters to
service chiefs. If the apparent intent of the 1964
change is realized, the effective coordinating powers of the prefects will be considerable.

These particular provisions do not apply to
the Paris Region. Because it is a vast urban region containing several départements, the coordinating mechanisms for Paris are being concentrated at
the regional rather than departmental level.

Also, as agent of the national government, the
prefect exercises executive police power, which
entails responsibility for maintaining public order,
morality, and hygiene and has authority to issue
regulations in these categories. Some newer national services, such as town and country planning, have
been put directly under his management. The prefect is generally responsible for guarding the
interests of the government in the département and
seeing that national law is carried out. Finally,
he is supervisor of municipal governments.

These activities of the prefect are on behalf
of the national government and are subject to
superior hierarchical authority exercised by the
minister of interior, who can overrule decisions of
the prefect. The prefect plays another role as
executive agent of the departmental council, however. In this capacity, he prepares and directs
implementation of that council's budget, legally
represents the département, makes proposals to the
council, and executes council decisions. In these
cases, the minister does not have direct hierarchical control over his decisions.

The prefect is thus a melting pot of conflict-
ing loyalties: agent of the département council,
adviser to mayors, and advocate of local interests

to the national government while agent of that
government. Strong tradition and the caliber of
the men holding the post account for the fact that
prefects do frequently fulfill these divergent roles
and succeed in maintaining the confidence of both
local authorities and the central government while
being an arbiter and major channel of communication
between them.

During periods of ministerial instability in
the Fourth Republic, Seine prefects have exercised
far-reaching powers. Whereas other départements
are headed by a single prefect, there are two prefects of Seine: a general prefect (prefect of the
Seine) and a prefect of police. The latter has the
normal police powers of the prefect, and the former,
the remainder of the prefectoral functions. As
there are no municipal police in Paris, the prefect
of police is in charge of all police functions in
the city and most of them in the rest of Seine,
where municipal police share in minor police functions. Housing and sanitary-code administration,
cemeteries, markets regulation, and fire services
also fall within his domain.

While the prefect of the Seine has had greater powers over the councils of Paris and of Seine
than is normal (restricted devolution), greater
powers are reserved also to the central authorities
vis-à-vis this prefect (restricted deconcentration).
These differences are based on the special structure
of required approvals for specified decisions in
Seine. The prefect of Seine has some original
powers that in other départements belong to the
councils (for example, approval of property assess-
ments, normally a function of council committees).
And central-government approval is required for
some decisions of the Seine council, which elsewhere would be validated by prefectoral approval.

The most notable distinguishing feature of
administration in Seine is, however, the third
dimension to the role of the prefects, for the City

of Paris has no mayor. The prefects of Seine have
the powers and functions of a mayor with respect to
the City of Paris. The general prefect calls the
city council and directs the city services (generally consolidated with departmental services). This
is the greatest source of controversy between the
advocates of greater independence for Paris and the
defenders of the central powers.

The administrative structure under direction of
a prefect is highly complex, including some field
services of the ministries, all of the staff of the
departmental council, and--in Seine--the staff of
the City of Paris. The prefects' immediate subor-
dinates include the subprefects, one stationed in
each arrondissement (a solely administrative territorial division that includes, on an average, 150
communes)*; the secretary-general--a professional
chief administrative officer in the département and
the prefect's right-hand man; and the cabinet chief --
a junior member of the prefectoral corps who functions as personal staff assistant. The subprefect
is the immediate supervisor and adviser of the
mayors in his arrondissement. The prefect of Seine
has three secretaries-general, one of whom operates
as commissioner of construction and urbanism. The
other two have assigned responsibilities over particular divisions and services.

The operating staffs in Seine-et-Oise and Seine-
et-Marne have been organized under five and four
general divisions, respectively, plus services that

---

*Seine is an exception in this respect; it con-
tains no regular arrondissements. There are twenty
subdivisions within the City of Paris called arrondissements. The town halls situated in each, headed
by an officer subordinate to the prefect, handle
vital statistics, registration, census, and other
routine tasks dealing directly with the public.

enjoy somewhat greater autonomy. Some of the latter
are branches of national services (operating under
ministerial budgets but attached to the prefecture),
and some are properly departmental services (operating under the budget voted by the council). Each
service is headed by a technical chief. The general
divisions are subdivided into bureaux. To take an
example, the departmental public-housing authority
is a departmental service that constructs and manages public housing under the council budget (with
national aid); there is also a housing bureau in the
division of communal and departmental administration,
which undertakes policy work relating to all public
housing, municipal and departmental, in the département.

The administration of Seine is larger and more
complex. The major general divisions include general inspection of services; cabinet staff; council
secretariat; general and departmental affairs; municipal affairs; social affairs; commerce and industry; fine arts and architecture; finance; housing;
personnel; transport; bridges and roads; urbanism;
and technical services (including a central project
preparation division and directorates of roads,
water and sanitation, public works, bridges). In
addition, the national services of health, population, and veterans are under partial control of the
prefect of Seine. Here in the Seine prefecture
beats the heart of administrative action for the
urban center.

This structure encompasses the integration of
national and local services in Seine. A single
technical service (water, or bridges and roads, for
example) under the direction of the prefect may
function as a city, departmental, or national service, depending on the activity involved and distinguishable by the budget to which the activity is
charged. Hence only part of the prefecture service
activities are subject to the powers and chargeable
to the budget of the council of Seine.

The Role of the Departmental Council

The council is the representative body of each département. Its membership is directly elected for a six-year term from electoral districts (cantons). Candidate lists are presented by political parties and coalitions of parties. The Seine council has been unique in that it included the ninety members of the Paris City Council, plus sixty members elected in the normal manner from the suburban areas of Seine.

The departmental councils have exerted considerable political influence on the national government from time to time. By tradition and out of political wisdom, the government and various ministers seek the opinion of the departmental councils on many matters of concern to the département. The councils vote the departmental budget, which is prepared by the prefect on the basis of submissions from prefecture services, and have general powers to legislate on matters of departmental interest. In addition to their own activities, the councils finance and regulate certain services on behalf of the national government, most notably, public-assistance programs. Their powers of decision are more definitive than those of municipal councils. Most council decisions (except in Seine) are self-executing. The budget is, however, subject to approval by the minister of interior and (by tradition) by the minister of finance.

All councils except that of Seine are empowered to appoint from their own membership a departmental commission that sits between the two regular sessions per year of the full council to supervise administration, check the prefect's monthly accounts, and exercise limited delegated authority. (The effective powers of the commission are kept in check by the proviso that councilmen who are also members of Parliament or the mayor of the departmental capital may not serve on it.)

The Seine council is larger and formally

weaker than other departmental councils. It can
only legislate on matters specifically enumerated
by special law, all other decision-making powers
being vested in the prefects. Its capacity to oversee departmental administration is limited by the
fact that it is not represented by a commission between its four ordinary sessions each year. Its
role with respect to the City of Paris is consultative only, while the other departmental councils
have certain supervisory powers over municipal
council activities.

Nevertheless, the budget of the departmental
council of Seine is the basis for a wide range of
urban services, including public-health and welfare
institutions and programs (for which budgetary receipts include statutory contributions from the
national government, the Paris City Council, and
other communes); public works (highways, bridges,
public buildings, etc.); public-housing programs;
sewage treatment and disposal; transit (the département covers a portion of the deficit of the
regional transit authority); and certain secondary
schools.

The officers of the Seine council include a
president, who presides over its meetings; four
vice-presidents; and four secretaries and a full-
time managing officer (syndic), who together form
the secretariat of the council to conduct its business between sessions. Six permanent committees
study problems and proposals and advise the full
council. Their concerns are departmental property,
roads and highways, water and sanitation; assistance,
child protection and public hygiene; education and
training, fine arts; finance, taxes and accounts;
land use, housing, and technical works; and police
and prisons. Each committee elects a permanent
secretary and performs the major portion of the
council work in its particular field. In addition,
a budget committee is comprised of the full finance
committee and representatives of the other committees.

Prior to 1961, Seine council decisions required higher approval unless specifically mentioned in
law as being self-executing. This principle was
modified by decree of January, 1961 (a "decentralization" decree), which enumerated those decisions
requiring higher approval and expanded the subjects
on which the prefect must consult the council.
Examples of decisions of the council that would have
legal force without approval are management of department property, acceptance of gifts, and classification of departmental roads. Some enumerated
decisions are self-executing only if they confirm
prefectoral proposals or are not vetoed by the prefect. All significant decisions on budgetary and
financial matters still require explicit government
approval (by prefect, ministerial order, government
decree, or law, depending on the case).

THE PARIS CITY COUNCIL

The most notable aspect of government of the
City of Paris is that it does not have its own
administrative structure. The two prefects of
Seine are its chief executives, and city administrative agencies are organized within the prefecture under the management of the prefects and the
departmental secretaries-general. Some municipal
services are suborganizations within the prefecture;
some are integrated with departmental services, in
which case the particular activity and budgetary
charge determines whether the municipal or departmental council is the controlling local authority.

The ninety-member city council is directly
elected from fourteen districts. Beginning with
the local elections of 1965, the total number of
seats for each of these districts goes to the party
list that receives the majority of popular votes in
the district (or a plurality if a second vote is
necessary).

Prior to the January, 1961, decentralization
measures, the Paris council was essentially

consultative. It was authorized to consider only
enumerated subjects and of these could make binding
decisions only on minor ones. On the other hand,
the prefect and national government were in the
habit of consulting it. The decrees of January,
1961, expanded the lists of matters on which it
could deliberate and of those on which the prefect
is required to consult it. Examples of additions
to the former category are contracts for the undertaking of commercial or industrial services, and
fixing priorities for capital projects. Additions
to the second category include disposition of city
property and authorization of capital works by
Assistance Publique, a special health and welfare
authority in the city.

The decrees also made all the council's decisions that do not explicitly require approval self-
executing (as they did also for Seine council decisions). While the Paris council was endowed by
these measures with greater powers of initiation,
it is nevertheless true that most major decisions
are still subject to higher approval in one of
several forms.*

The decrees of 1961 included deconcentration
measures (in addition to the decentralization measures), essentially simplifying approval procedures
for city council measures, shifting some approval
powers downward from the Council of State to ministers and others from ministers to the prefect. Most
decisions are now subject to approval by the prefect or the ministers of interior and finance.
Moreover, these measures rendered failure to reject
tantamount to approval after the lapse of three or
six months, depending on the subject. This is an

---

*Examples of self-executing decisions of the
council are those concerning management of communal
property, levy of certain local taxes as authorized
by national fiscal law, municipal guarantees of
loans to public-housing organizations and land-
credit institutions in accordance with national
regulations, and parking regulations.

important change because many projects died quiet
deaths in the past when departmental and municipal
loans and projects contingent on them were simply
shelved at some point in the central-government
approval process.

Examples of decisions of the Paris City Council requiring government approval by various authorities are as follows:

1. By national law: Creation of new taxes.

2. By decree of the Council of State: Authorization of loans with amortization periods over
thirty years; authorization of loans for works included in the general land-use plan for the region;
alteration of public-street plans if public investigation recommends against it; organization of a
public authority for commercial or industrial activity if the structure does not conform to alternatives defined by law.

3. By decree of the prime minister: The city
budgets and supplementary credits; municipal
accounts; capital programs.

4. By ministerial or interministerial order:
Creation of a public authority for undertaking a
public service (such as construction of low-rent
housing) if it is not structured according to
legally defined alternatives; authorization of loans
with amortization periods of less than thirty years.

5. By prefect of the Seine: Alteration or
establishment of street plans if public investigation does not yield a negative recommendation;
creation of major markets; transactions involving
more than 5 million francs; creation of an intermunicipal special district.

Like the Seine council, the Paris City Council
can pass resolutions and pose to the prefect questions on local matters that must be answered.

Of the officers of the Paris council, which
parallel those of the Seine council, the president
represents the city ceremonially. While only the
full council has formal powers, the committees do
the bulk of its work.* In recent years, the chairman of the budget committee has been the most out-
spoken political representative of the city and
advocate of expanded local powers.

The city council budget partially or wholly
supports public-health and welfare services, several
public-housing and urban-redevelopment programs,
streets and roads, water supply and sanitation,
school construction and maintenance, transit deficits, and police and fire services.

In summary, then, the formal powers of the
Paris and Seine councils over urban administration
have been and still are quite limited. Within the
government structure, however, these bodies have
influence, and the fact that they must vote the
budgets covering most urban government activities
demonstrates that the government must contend with
their opinions.

THE COMMUNES IN THE REGION

In addition to the City of Paris, there are
1,304 municipal governments or communes in the Paris
Region, with populations ranging from 85,000 to less

---

*These include six permanent committees: fin-
ances, taxes, public utilities and municipal services; general administration, policemen, firemen,
and property; streets; education and fine arts; public assistance and municipal loan bank; waters, sew-
ers, and hygiene. In addition, there are a special
budget committee, mixed committees with the Seine
council, and three special study committees.

than 2,000. They are subject to the general law on communal organization and powers that is uniform for all communes, into which all of France is divided.*

Chart 3 represents communal government schematically. The official institutions of the communes are the council and the mayor. Local councils are directly elected every six years on the basis of proportional representation in communes with less than 30,000 inhabitants, and by majority list in communes with over 30,000. The mayor and assistant mayors are elected by the council from its own membership.

The councils (of from nine to thirty-seven members), which meet at least four times a year, are the legislative bodies of the communes, responsible for regulating communal affairs, voting the budget, and organizing local administration. They are generally empowered to deal with the affairs of the community and to provide local services. Their decisions have the force of law except those enumerated that are subject to higher approval or other restriction. Obligatory public services as defined by national law and regulations must be provided by the councils. The uniform system of local government, however, embodies some necessary flexibility by scaling the cluster of mandatory services and the magnitude of obligatory expenditure to the size of the commune. Obligatory expenditure includes, for most local units, maintenance of official buildings and cemeteries, salaries and pensions of communal employees, construction and maintenance of primary schools, specific public-assistance activities, rural police, contribution to fire services, and debt service. If communal councils fail to

---

*There are some exceptions to application of general law to communes in Seine, notably that local police are controlled by the prefect of police.

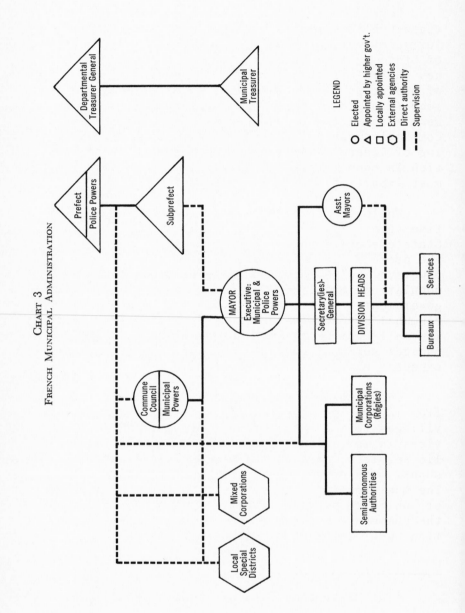

CHART 3
FRENCH MUNICIPAL ADMINISTRATION

Departmental Treasurer General

Municipal Treasurer

Prefect Police Powers

Subprefect

Commune Council Municipal Powers

MAYOR Executive: Municipal & Police Powers

Asst. Mayors

Secretary(ies)-General

DIVISION HEADS

Bureaux

Services

Municipal Corporations (Régies)

Semiautonomous Authorities

Mixed Corporations

Local Special Districts

LEGEND

O Elected
△ Appointed by higher gov't.
□ Locally appointed
⬡ External agencies
| Direct authority
¦ Supervision

48

vote funds for obligatory purposes, the prefect
can inscribe such charges into their budgets.

Beyond these matters, local councils can ini-
tiate public services and programs within their
jurisdictions, although permission of the Council
of State must be given for the establishment of a
new public-utility service in order that the issue
of public purpose may be determined. Decisions that
require approval of the supervising higher-government authorities include acquisition or sale of
communal property, alteration of street plans, raising of loans, establishment of markets, participation in commercial or industrial enterprises, and,
most important, establishment of the budget.

While the mayor's major role is that of executive agent of the council, he is also part of the
State executive under the hierarchical control of
the prefect and subprefect for exercise of police
powers,* and performance of some duties (such as
population registrations) assigned by the national
government. In the first capacity, he prepares
communal budgets (with the help of the subprefect
in small communes), executes council decisions,
manages communal property and administration, and
executes contracts for the commune.

The local administrative establishment varies
from mayor-cum-schoolteacher and rural policeman in
villages to a complex organization of public services in large towns in the region. Communal public services, such as bus transport and water
supply, are organized in three different ways.
The municipality may operate them through an agency
or public corporation (régie) wholly controlled by
the commune; they may be run by a private corporation under contract or concession from the local

---

*The mayor can issue ordinances relating to
public health, safety, and morality under the executive police power, but these are subject to nullification by the subprefect or prefect.

authority; or they may be undertaken by a mixed corporation (société d'economie mixte) in which communal and private capital participate. In all cases, a bureau within the general divisions of communal administration regulates and supervises such services.

In the large towns, the mayor is assisted by several assistant mayors and by one or more permanent secretaries-general, who are professional administrative managers of all or several divisions of communal administration. The assistant mayors often have responsibility to oversee specified divisions so that the work of the secretary-general is checked by a politically responsible official.

Communal personnel are locally recruited and paid but are regulated by national law as to categories, pay, rights and duties, and--in communes over 10,000--qualifications. A secretary-general, for example, of communes over 20,000 population must have a university degree, be over thirty-five years of age, and be recruited by examination. As the paid head of communal administration, or part of it, he is responsible directly to the mayor and is the immediate superior of division heads.

There have been no significant changes in the boundaries of the communes since the nineteenth century and earlier, when they were designed for rural settlement patterns. While the Council of State and the minister of interior are empowered to alter them, the major adjustment of local administration to urbanization in France has been the development of special authorities and intercommunal special districts (syndicats).* Intercommunal

---

*Neither annexation nor incorporation as practiced in the United States is relevant in France today, since all territory is part of an established commune. Loyalties to these units are strong, and the national government has been far more inclined

régies, for example, are public corporations managed
by a board grouping representatives of several participating councils. The prefect appoints one
fourth of the board, as well as the director and
fiscal manager of intercommunal régies, which are
generally single-purpose. The mixed corporation
allows one or several local governments to participate with banks and private interests in undertakings of a commercial or industrial nature. The public participation must by law be between 50 and 65
per cent. Urban redevelopment and construction are
undertaken through such companies in the city and
in several other parts of the region. A complex
body of regulations and administrative case law
controls the structure and operations (e.g., contract letting)of mixed corporations to protect public interests and revenue.

    Special districts and joint planning boards
are also authorized. The most frequently used is
the special district, entitled a syndicat intercommunal. In the Paris Region, these provide water,
gas, electricity, and undertaking services. They
have also participated in development projects in
the suburban portions of the region. Such special
districts can manage régies or other types of operating agencies to provide any services that are within the powers of the councils forming them. While
they may be multipurpose (the government is encour-
aging use of this form), most are established to
provide a single service. This type of special
district is formed by agreement of all local councils involved and order of the prefect, or by
agreement of two thirds of the participating local
councils representing over half the constituent
population and order of the minister of interior.

---

to stimulate creation of joint or umbrella special
jurisdictions than to tinker with communal boundaries. However, prior to 1861, the boundaries of the
City of Paris were extended several times.

It is managed by a board including two representatives of each participating council (one of whom is
usually the mayor) and financed annually by these
councils. They often receive, however, capital
grants from départements and the national government for major public works.

The pressures on municipal structures of
rapidly increasing demands for public services,
particularly in the fast-growing portions of the
Paris agglomération, have nevertheless not been
adequately met by these devices in the opinion of
the government. The possibilities of amalgamating
communes into units of from 25,000 to 100,000 population are under study. In the meantime, the reorganization measures of 1964 and 1966 affecting
departmental and regional institutions open the
door to another approach: increasing transfers of
public-service responsibility to these levels.

REORGANIZATION OF THE DEPARTEMENTS
IN THE PARIS REGION

The cumbersome administrative machinery of the
départements in the Paris Region (particularly
Seine), resulting from postwar expansion of activities within traditional structures as well as understaffing, and Communist and socialist strength in
Seine underlie the current reorganization. In general terms, Seine and Seine-et-Oise are being subdivided, each into three départements, none of which
include the City of Paris. Seine-et-Marne will re-
tain approximately its previous boundaries. The
Paris Region will thus include seven départements*
and the City of Paris, which becomes a unique entity in French administration--both commune and
département. The city council is unchanged in

---

*Hauts-de-Seine, Seine-St. Denis, Val-de-Marne,
Essonne, Yvelines, Val-d'Oise, and Seine-et-Marne.
See Map 2.

MAP 2
REORGANIZATION OF THE *Départements*
OF THE REGION, 1964-68

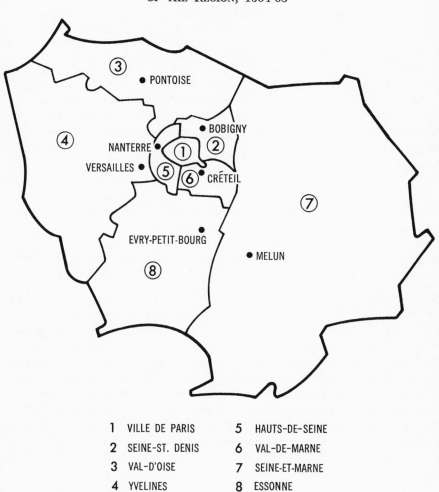

1	VILLE DE PARIS	5	HAUTS-DE-SEINE
2	SEINE-ST. DENIS	6	VAL-DE-MARNE
3	VAL-D'OISE	7	SEINE-ET-MARNE
4	YVELINES	8	ESSONNE

Source: Délégation Générale au District de la Région de Paris.

composition but assumes the powers previously exercised within the city by the Seine council as well.
This considerably simplifies administration for the
city, as many services and projects previously depended upon budgetary decisions and procedures by
both councils.

At the same time, the new organization separates
agencies of city administration from those of the
immediate industrial and residential suburbs, tantamount to what might be viewed as administrative de-
consolidation. The chief executives for the city
will still be a government-appointed prefect and a
prefect of police. The first has jurisdiction in
the City of Paris alone. The powers of the second
also extend to the three départements carved out of
Seine.* Thus, police powers are the retained category of consolidated administration for the city and
its environs.

In other respects, the general laws on departmental administration in France hereafter apply to
the seven new départements. The special regime of
Seine is eliminated; but special treatment for Paris
in general is not. In fact, that concept is strengthened by the modifications in central city government and, above all, development of regional institutions.

The political ramifications of these changes
are still vague. While the powerful left in Seine
is fragmented, the redrawing of boundaries creates
at least one new département (Seine-St. Denis) that
has a sure Communist majority. The powers of the
national government, the prefect of police, as well
as emerging regional authorities, however, will form
significant constraints on the latitude for action
by opposition-controlled jurisdictions.

---

*These three départements actually include some
thirty-five communes that were outside of Seine, as
well as its former territory.

The proposal for departmental reorganization was prepared by the government (with direct involvement of the President) and remained secret until its submission to Parliament. Organized and vigorous opposition to it did not form.

Such fragmentation of units of urban administration runs counter to most trends of urban reform in other nations where economies of scale, duplications in staffs and programs, and disjointed programming are recognized problems. They are recognized also in Paris, however. The reductions in departmental size must be understood in light of evidence of accompanying modifications in store for commune and regional institutions. There is gradually emerging an image of a three-tier pattern of administration for Paris composed of amalgamated and enlarged communes, the new départements, and emerging regional institutions.[1] Three tiers are viewed as a structure that synthesizes values historically attributed to small jurisdictions in France and desires to create a structural base for large-scale urban administration and regional services. While no steps have been taken to alter the communes (and evidence of government intent in this respect is scanty), movement on the regional front has been considerable since 1960.

REGIONAL INSTITUTIONS

In 1961, the government created the District of the Paris Region, the territorial jurisdiction for which was the three départements of the region. Its executive, the delegate general, rapidly became a dynamic leader who wore several other hats: staff adviser to the prime minister, secretary to the Interministerial Committee for the Paris Region, president of a separate planning and research body (Institut d'Aménagement et d'Urbanisme de la Région Parisienne, or IAURP). In 1966, he became the regional prefect of Paris, a new post that replaced that of delegate general and added to his powers and duties.

The Paris District is neither a general government tier nor a regional council, but a special district with generalized functions to plan, to coordinate, to study and advise, and to stimulate
planned development by selective investment of
limited funds. Like many elements in French administration, the district has a dual nature: It consists of a council of local representatives and a
State-appointed executive; it has specific powers
enumerated in the creating legislation and powers
delegated to it from time to time by the national
government; the executive is both agent of the council and agent of central authority; and finally, it
is a creature of the national government and yet
exerts pressure within that government for the
interests of the region.

The history of regionalism has been spotty in
France. The unpopularity of the Vichy super-pre-
fects is still remembered,* but the need for some
machinery on a regional scale, particularly in conjunction with national and urban planning, has been
recognized by the government for some time. The de
Gaulle government has made several steps toward
developing such machinery throughout France. The
General Planning Commission is undertaking regionalization of four-year national economic plans.
Regional committees that bring together officials
in départements are being utilized in an advisory
capacity in the planning stages and as voluntary
coordinating mechanisms for plan implementation.
For each of twenty regions of France, one of the
prefects of a département within it has been appointed coordinating prefect for plan implementation.
This degree of regional organization, however, was
not considered adequate for the Paris Region, which

─────────────────────────

*Since the war, the IGAME, special regional
inspectors attached to the interior ministry, have
been used mainly for purposes of law, order, and
emergency action.

is a huge urban complex, as well as an economic region. The district established for the Paris Region
incorporates a stronger regional planning role and
was designed to grow also into an administrative
agency.

The history of the district's creation demonstrates the informal powers of local authorities.
By the late 1950's, the lag in public investment
and the inadequacies in transport, housing, schools,
hospitals, and other urban services in the region
had raised the interests of experts and government
officials in administrative change. The lag was
partly attributable to the complexity of the system
of government and lack of focused power in the region and partly to national policy to channel investment elsewhere in France. * Beginning in 1955, a
group of local politicians and scholars sponsored a
series of studies[2] calling for establishment of a
national secretariat for the Paris Region that would
consolidate and simplify the exercise of national
powers and activities relative to the region, the
complexities and fragmentation of which have frus-
trated local Paris authorities for some years.

Instead, an ordinance of February 4, 1959,
established the Paris District with direct administrative powers--without altering the organization
of ministerial activities. Strong opposition developed within local governments (most of the population remained unconcerned with or unaware of the
controversy), mainly in the city and the parts of
Seine that feared further central domination, particularly in light of the history of the government
policy to slow the growth of the region. This ver-
sion of the district collapsed when the city and
several suburban communes refused to name representatives to the council. (In addition, the

----

*Until the national government modified its
emphasis on restraint of investment in the Paris
area, it showed little interest in attacking the
administrative problems of the region.

constitutionality of other provisions of the ordinance was called into question.) The government submitted a bill to Parliament in July, 1960, that gave
the district urban planning and investment functions
and eliminated its immediate administrative role.
This was finally passed in August, 1961, and the
district was organized in the same year.

The district is legally a public agency with an
autonomous budget. While its primary function is
that of research and planning--particularly with
respect to problems of land use and urban development, provision of public capital facilities, and
organization of public services--it also has a financial role: to aid local units (public and semipublic corporations included) with grants and loans
for development projects. It is empowered to finance totally public works with the agreement of the
public establishments involved. Finally, and only
with the agreement of all local bodies concerned, it
can implement programs, manage services, and undertake works of regional interest or contract with
local bodies to provide services. The district has
not undertaken any implementing activities under the
last rubric. Legally, it could assume operating
responsibilities without local approval by delegation
of the government--by decree of the Council of Ministers on opinion of the Council of State. No substantial administrative powers of traditional units of
government have been transferred to the district.

The district council (literally "board of directors"), which is legally the governing board of
the organization, includes twenty-eight members all
of whom are members of commune and département
councils in the region. It is to be enlarged to
provide for representation from the new départements. Half are elected by the councils and half
are appointed by joint order of the prime minister
and minister of interior. Geographically, the
representation has been distributed as follows:
City of Paris, 8 (4 appointed, 4 elected); département and suburban communes of Seine, 8 (4 appointed,
4 elected); département and communes of Seine-et-

Oise, 8 (4 appointed, 4 elected); département and
communes of Seine-et-Marne, 4 (2 appointed, 2 elected).

The law creating the district grants the council powers of decision, on advice of specialized
study sections, over the district budget* and advisory powers over regional plans. In actuality, the
nationally appointed executive--the delegate gen-
eral--has become the center of district power and
activity.** Internal organization of the council
is set forth by government decree. Sessions are
called by the district executive, who must do so if
requested by two thirds of the membership. The budget committee has been the most important channel
of communication between the council and the delegate general. There are three additional permanent

---

*The administrative reform law of June, 1964,
however, empowers the ministers of interior and
finance to inscribe expenditure for "priority projects" into the district budget if the council fails
to vote such expenditure twice consecutively. This
echoes the power of supervising authorities to inscribe obligatory expenditure in general local budgets.

**M. Delouvrier, the first and current execu-
tive of the district, has maintained close relationships with national authorities as well as with
local representatives. He has a sizable staff in
the district organization (some 120 employees), and
the research of the IUARP is at his disposal. The
council, on the other hand, has no independent technical staff to evaluate plans and investment proposals. Service on it is a part-time activity for
the members. And, as projects included in the district budget generally depend upon additional
national government financing, the council is dependent upon the executive and his staff for devising
plans and including programs that will receive
government approval.

committees--plans, programs, and projects; administration and finances; and works--the heads of which
are represented on the budget committee.

Six specialized study sections were included in
the district structure by statute in order to broad-
en participation. (Parliamentary deputies from the
region who were not also local councilmen supported
this provision.) Each of these sections, which examine proposals before the council, has no more than
twelve members chosen by the council on nomination
by the delegate general from among parliamentary,
local, and département council representatives from
the area.

To further broaden participation in formulating
regional goals, the delegate general successfully
urged the district council to establish the Economic
and Social Consultative Committee. Its fifty-five
members appointed by the council are drawn from
trade unions, employees associations, professional
and other social and economic community groups.
Creation of this body, together with a public-
relations campaign and conduct of opinion polls in
the region, represent determined efforts by the district in its early years to create a public consciousness of the region as a unit and to plow the
ground for growth of consensus on major urban problems and solutions.

The delegate general to the Paris District was
appointed by decree of the Council of Ministers and
was responsible directly to the prime minister. He
is charged with proposing policy to the government
on planning, investment, and development for the
Paris Region, as well as with acting as executive
of the district. He also exercises some functions
that have been delegated by ministers of the government. The district staff of approximately 120
under his direction consists of civil servants de-
puted from various regular national services to
serve the district and paid by the national government. It is organized in three permanent sections: general research, finances, and programs.

There are also three agencies exterior to the district organization that work closely with it:

1. The Paris Region Institute for Planning and Urbanism (IUARP), a public corporation created by the Ministry of Construction and engaged in research for the district on urban planning problems and land use;

2. The Town and Country Planning Service for the Paris Region (Service d'Aménagement de la Région Parisienne, or SARP), a regional branch of the planning and research service of the Ministry of Construction;

3. The Service for Coordination of Public Facilities Development in the Paris Region, of the Ministry of Construction.

The Institute, of which the delegate general to the district is president, is in fact undertaking most of the basic studies needed for the district plans and is financed in large part from the district budget.

As is common in major French administrations, there are a cabinet, chief of cabinet, and secretary general serving the delegate general. In addition, he has established fourteen work groups, each organized around a functional area of district interest as defined by the council. These groups, which include technical experts from ministries and the département agencies in the region, help to elaborate plan proposals and to stimulate coordinated activity.

The authorized financial resources of the district include: (1) taxes, fees, and contributions from or relating to services it may provide; (2) voluntary contributions from départements, communes, or public authorities; (3) a special taxe d'équipement; and (4) financial participation by other units in projects it may undertake. The

taxe d'équipement levied in the region by the national government is in fact its only major source of
current revenue to date. It is an addition to local
taxes levied against legal persons and is apportioned among the communes in the region by formula related to local tax base. The minimum total under
law of 170 million francs annually (some $37 million)
was the district budget in its early years. By
1966, its budget had risen to 200 million francs.

In 1965, however, the district began to borrow
capital on its own account and has since under-
written some local borrowing. This was a controversial exercise of power, not explicitly authorized in
the creating statutes and decrees, but deduced by
district authorities from the body's legal status
of "financial autonomy." Until 1965 the traditional stance of the national government against large-
scale borrowing by local authorities in Paris was
projected onto the district. The borrowing proposal included in the district budget of 1964 was,
however, not eliminated but reduced by the minister
of finance. A rider to the national budget law in
1965 formally recognized the district's borrowing
powers.

Fundamentally, the district is designed to
help coordinate policy and activity among units of
local government, as well as to help develop comprehensive national government policies for the
region to be applied through the decisions of the
government, the Interministerial Committee for the
Paris Region, and the activities of various ministries. On the local level, it depends on the voluntary cooperation of local authorities, who view
it with reservations. Its effect on national-
government behavior is highly dependent upon the
influence of the delegate general and support from
the prime minister or President to wrest concerted
action from rival ministries. The present delegate general is a strong and respected personality
who has increased government awareness of the
region's needs. Nevertheless, while he is the

major policy proposer to the government, he is first
dependent on it for definition of the short- and long-
term policy alternatives within which he has to work.
One of his assistants, Michel Piquard, has summed up
the district's position thus: "...as we have very
little authority, we must act by gaining influence."

In 1966, regional organization of national services in Paris was strengthened by statute and executive decrees that established the post of regional
prefect for Paris. M. Paul Delouvier, formerly delegate general to the district, was appointed to the
position. The regional prefect is generally charged
with implementation of government policies on economic and urban development and supervision of regional branches of national services (except those directly responsible to central authorities such as
education and justice). For these purposes, he has
some authority over the service chiefs and departmental prefects in the region. In addition, he
assumes the powers formerly vested in the delegate
general as executive of the district and adviser to
the government. He is, however, henceforth responsible to the minister of interior rather than to the
prime minister, as was the delegate general. His
immediate staff is to be of high caliber. Two
secretaries-general and a staff aid (cabinet chief)
serving him are to come from the prefectoral corps
of the civil service.

The regional prefect nevertheless does not have
the regular executive powers of a departmental prefect. He exercises neither approval over actions
by commune or departmental councils nor executive
police powers. His primary tasks are coordination
for plan implementation, particularly of the current investments in the regional transit system and
land development. Finally, he will call meetings
of a new interministerial committee on land-use
planning and preside over a committee of regional
administration, which will bring together departmental prefects and regional service chiefs.

Most public services continue to be provided

by ministerial service branches, departmental and
communal councils, and existing special authorities
in the region, such as the regional transit authority. The prospects for regionalization of finance
and management in water supply, sewage disposal,
and public utilities are raised, however, by recent statutory authorization to the government to
create public establishments to handle regional services (only in the Paris Region) by central decision
alone.

There has been some consideration given to
creation of elected regional councils with broad
powers in France in general and Paris in particular.
This step is viewed with some caution by many in the
government who fear it might federalize France. The
pressures for metropolitan-wide organization in Paris
may thus ultimately produce a host of special authorities at the regional level, which would certainly
tax the coordinating capacities of the regional prefect. The government's attitude toward Paris and
the potential power of the region mitigates against
creation, in the alternative, of a general regional
government with its own taxes, budget, and statutory powers.

Notes to Chapter 2

1. See, for example, Michel Piquard, "Organization and Planning of the Paris Region," Public
Administration, 43 (Winter, 1965), 383-93.

2. Jean Legaret, et al., Le Statut de Paris
(Paris: Librarie Générale de Droit et de Juris-
prudence, 1956-58), 2 vols.

CHAPTER **3** INTERGOVERNMENTAL
RELATIONSHIPS

In view of the integral nature of the administrative system for the Paris Region, a purist would
entitle this chapter "Intragovernmental Relationships," but the local, departmental, and central
government units have separate budgets, and it is
enough to keep in mind that they are essentially
subsystems of a single administrative system.

Special treatment of the City of Paris, the
Département of Seine, and recently the Paris Region
in most respects has resulted in intensified centralization of powers and control. Though centralized, however, authority to govern Paris is not concentrated. It is scattered among ministers, field
services, officers, and special authorities to the
extent that the major dimensions of coordination for
regional action are the concentration of central control and coordination of national authorities in the
region.

Intergovernmental relationships in the region
are extremely complex, nonetheless. Every major
developmental activity requires complementary decisions from several units of government, many of
which have overlapping responsibilities. The
structure imposes a degree of inertia that only
strong leadership can pierce. Most of the reforms
since 1960 have been aimed at making administration
for Paris more responsive to leadership, more amen-
able to getting things done.

ALLOCATION OF POWERS

Responsibility for Various Public Services

The public services for the Paris urban area
are not neatly allocated among levels of government.
The intergovernmental organization is increasingly
specialized by process (such roles as budget-making,
planning, investment decisions, and direct operations) rather than by purpose (service areas such
as health and education). All levels of government
are involved in public housing, water supply, urban
transport, education, roads and highways, welfare
and health programs, and planning and redevelopment,
although the relative importance of national and
local authority roles in each functional cluster of
activity varies. At either end of the spectrum,
for example, education and police in the Paris Region can be identified as primarily national services,
while water supply and sewage are considered municipal services. Communes are responsible for building
and maintaining elementary schools, however, and
suburban communes have municipal police forces. On
the other hand, approval by the national government
of financial and technical aspects is essential in
all major improvements and extensions of water and
sewage systems.

In the French system as it operates in the
Paris Region, formal responsibility of a particular unit of government for a particular function
may only mean that the budget of that unit must
provide for the expenditures on that function, or
a certain proportion of them. Thus, while water
supply is a communal responsibility, in rural and
some suburban sections of the region the facilities
are actually operated by departmental services, the
costs of which are reimbursed to the département by
the commune. Responsibility for a public service,
even so precisely defined, is often shared by various units. Thus, while the prefect of police is
the formal and operating fire protection authority
for Paris and Seine, the city and suburban communes
make obligatory financial contributions to general

revenues for fire services, for which they receive
some offsetting national subsidies.

There are many examples of the interpenetra-
tion of various levels of government within service
areas: Curative public-health functions in the City
of Paris are provided by Assistance Publique, which
is legally a city agency but is closely controlled
by central authorities (its director is appointed by
the minister of public health and its board is headed by the prefect) and is jointly financed. For the
most part, parks and recreation facilities are provided by local authorities in the region; however,
major Paris parks are national, all local recrea-
tional programs are supervised by the departmental
divisions of youth and sports, and 50 per cent
national grants are available to communes for construction of new facilities. Major cultural facilities are national, but there are local libraries
that are subject to inspection by the Ministry of
Education.

While this mix of governments in each service
is a common trend in urban areas of other nations,
it is particularly intense in France because communes and départements have general powers. Under
law, the services they may provide are not precisely defined (although their powers are), nor is
special enabling legislation necessary for them to
undertake new activities. Any unit of government
can build public housing, hospitals, or swimming
pools. In a few cases, such as police and education, the permissible functions of local authorities are limited by law. In others, such as public
assistance, the general involvement of all levels
of government posed problems of duplication that
led to a more rational and explicit statutory
allocation of powers and costs. The costs of public-
assistance programs are now apportioned among
the national government, départements, and communes
by complex formula; most of the operating activities are undertaken by the département, but claims
for assistance are made through communal authorities.

The Roles of Government
at Various Levels

Although each level is involved in each major
service, some patterns can be discerned in the sort
of role played by various types of agencies. Initiation of services and service projects has fallen
predominantly to communal and departmental councils.
A reservoir, a local market, a departmental highway,
a new elementary school, an urban-renewal project--
each must be voted by the appropriate council.
Initiation powers are qualified in that the exercise of some of them are obligatory, of course; but
local governments in the urban area have gone far
beyond required levels of obligatory services. A
new participant in initiating projects is the district, whose role in this respect will undoubtedly
grow with its plans and budget; but thus far, it is
a cooperative initiator, stimulating projects that
must also be voted by local authorities.

Implementation (both management of services
and execution of projects) is primarily a communal
and departmental activity. More technical tasks,
such as highway and reservoir construction, are
undertaken by departmental or City of Paris services. Operation of water-supply facilities,
school maintenance, and hospital maintenance tend
to be undertaken by communes and the city. Département and commune agencies have implementing roles
in some strictly national services also. Mayors
must collect data for national censuses; départements must provide overhead facilities for field
services of the Ministry of Education; communes
must administer national unemployment assistance;
and Paris must maintain national highways. In
these cases, the local council plays little part
except to vote obligatory expenditure. It is true,
however, that field services of ministries do
undertake some direct operating functions in the
Paris Region: They hire and supervise teachers
directly, build national highways in the urban
region, issue certain construction permits, prepare certain town plans, and administer justice.

A second qualification is that special operating
authorities are increasingly utilized--such as the
intercommunal special districts in water supply,
mixed corporations in urban renewal, and a major
regional public authority in transport--that, because of the nature of central controls over them,
increase the involvement of the national government
in detailed implementation.

Finance for urban services in the Paris area
is highly fragmented among the national budget
(ministerial service budgets and special treasury
funds), departmental council budgets, and communal
budgets. The practices of joint governmental participation and required contributions of one or more
units of government for activities undertaken by
another, as in the case of Seine fire services and
regional transport-authority deficits, arise from
the allocation of financing responsibility by special laws, regulations, and agreements that vary
from case to case.

Major capital resources are provided at least
in part by the national government in almost every
urban service category. Commonly, the national
budget meets up to one third of the costs of capital projects in the region, and a higher proportion
for some.* In effect, while departements, communes,
and the city contribute to public-investment projects, the national government has a dominant role
in capital credit both through direct grants, loans,
and loan guarantees and through its tight control
over local borrowing powers.

It follows that major decision-making powers
in urban administration in the Paris Region rest
with the national authorities. By this is meant,

_____

*Of the governmental capital expenditure called
for in a four-year plan for the region, about 37 per
cent is allocated directly to the national government. See Table 3 in Chapter 4.

first, that every major urban-project proposal will
die unless favorable decision is taken by several
central authorities; and second, that the decisional
framework of local operations is predetermined by
national regulations and checked by ongoing supervision. When a local authority decides to do some-
thing, the administrative and technical aspects of
how it is to be done are determined in large part
by national law and regulations. This situation is
not entirely based on explicit and legally defined
allocation of powers. On the contrary, while the
decision-making powers of the Seine and Paris councils are restricted, other communes and départements
formally have powers to decide local matters. The
situation is derived from the limitations of local
finance, the systems of control--administrative,
technical, and financial--of local governments, the
cumulative effect of diverse national regulations,
and the regional impact and technical nature of most
current decisions and services in the urban complex
that intensify national involvement.

In addition, of course, the existing degree of
centralization of decision-making is based on the
effective political power of the present government and its attitude toward Paris. Nothing in the
formal administrative structure fully explains why
such detailed decisions as those on highway routing
and parking programs in the Paris area are taken in
the Council of Ministers or Interministerial Committee for the Paris Region.

These points will be illustrated more clearly
in the examination made in Chapter 5 of water
supply, public housing, education, and transportation programs in the region.

Financial Resources [1]

The national government controls the lion's
share of public finance resources in France and in
the Paris urban area. In 1960 and 1962, the pro-
portions of general operating revenues of the
national government, all départements, and all

communes in France were roughly 81 per cent, 7 per cent, and 12 per cent respectively. National-government expenditure has been growing fastest: Between 1950 and 1960 the national budget increased by over 300 per cent, while departmental and communal expenditures increased by 175 per cent and 145 per cent respectively. (The absolute growth rates are considerably distorted by inflation.)* At the same time, the magnitude of local expenditure in the Paris Region is high. The Département of Seine accounts for about 20 per cent of all departmental expenditure in France (having roughly 12 per cent of the nation's population); and the City of Paris accounts for over 20 per cent of all communal expenditure in France (having only about 6 per cent of the nation's population). About 20 per cent of the national government's investment expenditure** is made in the region as a whole, this proportion closely corresponding to that of regional-to-national population. While the national government is not favoring Paris, the per capita financial burdens on local budgets are higher in the region than elsewhere.

The national government invests directly the largest amount of public capital expenditure in the Paris Region, with the City of Paris supplying the second greatest amount of capital. In 1962, of total capital authorizations by government in the region, roughly 54 per cent was by the national government, 14 per cent by the City of Paris, 12 per cent by all other communes, 11 per cent by the three départements, 7 per cent by the district, and 2 per cent by local special districts and public corporations. The City of Paris accounted for two

---

*The gross general price index for 1958 to 1962 was: 166.9, 174.9, 179.4, 183.1, 188.1. The base of 100 is for the year 1949.

**The French budget format does not yield to breakdown of total national expenditure by region.

thirds of communal operating expenditure* in the
region.

Central-government dominance of financial resources appears even greater upon analysis of the
revenue sources from which local expenditures are
made. Local taxes accounted for about 45 per cent
of département operating revenues in the Paris Region in 1960. Some 25 per cent was derived from
national funds, and most of the remainder came from
contributions from national--and to some extent from
local--units for particular services, notably social
security and public assistance programs. About 17
per cent of communal operating revenues in the urban
area were from higher government funds (rembourse-
ments, subventions)in 1960 and 1962.

To carry the analysis of the allocation of financial power a step further, we find that the ability to increase fiscal resources is mainly in
national hands. Local taxes include "additional
centimes," transaction taxes, and various others
as authorized by national law. The first are
essentially taxes based on real and personal property and professions that are assessed and collected by the communes. The value of the "centime,"
or mill, is calculated as a percentage of the revenue the national government would have collected
if it levied the national property tax, which it
has not done since 1918. Prior to that year, for
every franc the citizen paid the State, he paid a
certain number of additional centimes to the commune and département. The number of additional centimes were and are voted by commune and département
councils.**

_____

*The terms "operating revenues" and "expendi-
tures" refer to the section de fonctionnement of
the budgets as opposed to the section d'investisse-
ment. Some capital transactions are included in the
former.

**The bases of assessment are fourfold: unim-

Reassessment of old property holdings is rare,
and the tax bears little relationship to inflation
or real wealth. To meet their needs, many local authorities levy tens of thousands of additional centimes. These taxes are declining in relative importance in the urban center, comprising less than 10
per cent of the total revenues of the City of Paris.

Becoming more important is the transaction tax
on services and wholesale and retail sales. This
is collected by the national government at a rate
that varies with the volume of the transaction to a
maximum of 2.75 per cent. Of the yield of this tax,
a national equalization fund absorbs 10 per cent.
Prior to 1964, another 35 per cent went into a Département of Seine equalization fund. The remainder
is returned to the commune of collection.

The equalization funds have failed to offset
substantially the inherent bias of this tax against
the rapidly growing suburban residential communes,
which must levy far higher rates of "additional centimes" in order to provide lower levels of public
services than communes in the urban center. This
prevailed in spite of the fact that Seine made contributions from its own equalization fund to Seine-
et-Oise and Seine-et-Marne.

In 1964, a regional equalization fund was
established, unique to the Paris Region, which re-
places the departmental fund. In addition, the
expenditure of the Paris District has been allocated to priority projects, including suburban water
supply for example, which somewhat increases public

---

proved real estate, improved real estate, personal
property, and patantes, which are license fees for
professionals and certain types of entrepreneurs
based on rental paid to do business. Only the last
has increased consistently since 1930. The Paris
District tax is allocated among communes according
to this combined tax base.

expenditure (mainly capital expenditure) in the suburban portions of the region.

Other local taxes include several miscellaneous
indirect taxes, a capital-gains tax on real estate,
and a tax on rents paid by tenants of houses and
apartments.

Over-all, the leeway within authorized taxes
for local authorities to increase their revenues by
their own decision is narrow. Local authorities
tend to cite the burden of national taxes as a limit
on their willingness to raise local taxes, in any
case. There is a substantial margin for increase,
however, in another category of local revenues--that
is, user charges for such services as water supply,
sewage disposal, and garbage collection. The restraints on local authorities in this respect are
political, as is common in all urban areas studied.

The limits of local financial power are a
function not only of the restrictions on their
revenue-raising powers but also of the increasing
burden of obligatory expenditures. Some 80 per
cent of the City of Paris' operating budget for
1964 and 75 per cent of it for 1963 consisted of
obligatory expenditures. Over 90 per cent of the
1964 and 1963 budgets of Seine was obligatory expenditure. As discretionary expenditures in these
units have remained fairly steady between 1960 and
1965, most of the increase in local revenue has
gone into obligatory categories (including such
arbitrary items as maintenance of national highways and courts).

Finally, local borrowing powers in the Paris
Region have been severely restricted. Local credit
transactions are subject to specific approvals, and
most local borrowing is from national banks and
credit institutions. From the end of World War II
to 1959, the City of Paris was not permitted to
borrow at all. Municipal and département councils
in the region voted numerous loans for projects
that were never approved. Since 1960, the borrowing

powers of Paris and Seine have been expanded; together, they were allowed to borrow 260 million
francs in 1964. In addition, limited borrowing by
the district has been authorized. The magnitude of
investment needs, however, is such by comparison
that national-government participation remains
essential in major projects.

National-capital participation is of two types:
direct investment in joint projects, which does not
show up on local investment budgets; and special
capital subsidies granted for particular projects
by individual ministries, which entail thorough
technical control. The latter accounted for about
22 per cent of communal investment budgets (including Paris) in the urban area in 1962.

INTERGOVERNMENTAL CONTROLS

Conceptually, it is useful to distinguish between the allocation of legal responsibility and
resources to local authorities in the urban region
and the controls maintained over their exercise of
such responsibilities and utilization of such resources.*

Insofar as prefects and mayors in France act
as agents of the State--in exercising executive

---

*This distinction has proved useful particu-
larly for comparative purposes and for analysis of
problems of urban administration. A hypothesis
that the fewer the statutory powers and resources
assigned to local authorities, the less intense the
operating controls over their activities, has proved
untrue on a cross-national basis. On the other
hand, a hypothesis that intensive and detailed
operating controls where few powers are initially
allocated to local authorities reach a point of
diminishing returns in terms of both administrative efficiency and effective policy control appears
to stand up.

police powers, for example--they are, like field
personnel of the central ministries, subject to
direct hierarchical control. Thus, in this category of executive action, mayors are subject to
direct orders or counterorders from subprefects or
prefects--and prefects, subject to the minister of
interior.

Over powers of decision granted to persons not
subject to hierarchical control--local councils and
their executive agents, and public authorities--a
general system of supervision comes into play:
tutelage or guardianship (tutelle). The supervising authority (autorité de tutelle) over communal
councils and mayors for most purposes is the prefect, through his subprefect, and for département
councils is the minister of interior. In Paris
and Seine, however, the minister of interior is
the direct supervising authority over local units
for numerous purposes. Tutelle à priori involves
granting of prior approval for local-government
action; tutelle à posteriori is the opportunity for
government nullification on legal grounds of local
action taken. Both methods of control are utilized in administrative and financial supervision.

Financial Supervision

While the apex of administrative supervision
is the Ministry of Interior, the tight network of
financial controls gives the Ministry of Finance
crucial powers in the Paris Region. It is influential in budget approval and directly involved in
audit and financial administration. The budgets of
communal and departmental councils must follow pre-
scribed form and must be balanced.

The budget of the City of Paris is subject to
approval by the prime minister. All other communal budgets are approved by the prefects, except
that approval by the minister of interior is required for communes of 80,000 or more inhabitants
(of which there are three in the Paris Region be-
sides the city) if accounts from past years show

deficits. Departmental budgets are subject to the
approval of the minister of interior. These authorities are empowered to eliminate or reduce revenue
and expenditure items for reasons of illegality or
lack of fiscal justification. They can only inscribe items of obligatory expenditure into a communal budget if the council has failed to do so.
The exercise of this authority is subject to appeal
to the Council of State, and it is, in fact, generally used with discretion to assure fiscal responsibility. Prefects and the minister are reluctant to
provoke political battles over budgets that may end
in the council's refusing to vote the budget at all.
Generally, they try to persuade councils to adjust
objectionable items in the budget before it comes
to a vote.

With respect to the City of Paris and Département of Seine, however, the minister of finance has
maintained particularly tight control over budgets
and expenditure, often through tacit agreement with
the minister of interior. This control has eased
slightly in recent years as the city and Seine
councils have been allowed to establish four-year
investment budgets financed through borrowing.

All public accounts are subject to audit by
the Court of Accounts, attached to the Ministry of
Finance, and to spot checks by the Finance Inspec-
torate, an agency within that ministry. Full investigation of irregularities or inefficiency can
be ordered on the basis of their findings.

The most significant aspect of financial
administration is, however, that day-to-day financial transactions of communes and départements, as
well as other ministries, are handled by officers
of the Ministry of Finance. The chief financial
officer in the département, the trésorier payeur
général, is the direct superior of treasurers of
the communes (receveurs municipaux) (see Chart 3).
While the mayor authorizes financial transactions
on behalf of the commune, the transfers are actually made by these finance field officers. The

trésorier payeur général has far exceeded the role
of legal fiscal supervision and has become in effect
a financial prefect who applies the financial and
economic policies of the Ministry of Finance and
supervises the exercise of budgetary approvals of
the prefect. Conflicts between the two departmental
officials must be settled between the Ministry of
Interior and Ministry of Finance within the central
government.

In addition to the regular system of financial
controls over local authorities, the Ministry of
Finance maintains control over the use of national
funds (special loan funds as well as ministerial
budget funds), which have become substantial portions of urban finance--particularly investment
finance--and over fiscal activities of public
corporations and special agencies such as the regional transport agency and the Paris District.
It can block national-government participation in
local capital projects, even though they are approved by the relevant technical ministry, on grounds
of fiscal policy and regulations.

To illustrate the finance ministry's power, in
1964 the District of the Paris Region (the executive of which was responsible directly to the prime
minister) voted a separate budget of receipts and
expenditure from long-term borrowing of 457 million
francs. The loan budget was to be utilized for
capital participation in construction of the transit express system by the regional transport agency.
The project had been the most important component
of the regional plan that had been approved by the
government. This budget was submitted by the district, with its regular budget, to the ministers of
finance and interior at the end of January. On
March 5, the directorates of budget and treasury
of the Ministry of Finance notified the district
council of objections: (1) the division of the
budget with a separate part for receipts and
expenditure from borrowing for a special purpose
was unacceptable; (2) total capital program

authorizations would have to be scaled down to necessary payment schedules as these became evident and
could not be designed on the long-term basis;
(3) some aspects of the district budget covered
matters requiring decision by the Economic and Social
Development Fund; (4) the scale of the district investment schedule was too high in light of government stabilization policy; (5) the magnitude of the
proposed borrowing was too great; and (6) the district tax would have to be raised to cover amortization.

The minister of interior concurred that a
separate borrowing program for the single project
was inadvisable. At the end of negotiations, the
district was authorized to float general equilibrium
loans only; all receipts and expenditures were to be
pooled and the district was to borrow only to the
extent of the deficit in each fiscal year. District
investment programs were reduced.

Administrative Supervision

General supervision of commune and département
councils (tutelle) is exercised for the most part by
the minister of interior and officers responsible to
him, although national regulations specify instances
of supervision over specialized local activities by
the technical ministries.

Approval of council decisions is the most
important aspect of administrative supervision in
the Paris Region. The approval procedures to which
most decisions of the City of Paris and Seine councils are subject were outlined in Chapter 2. These
approval powers are discretionary and amount to
control of policy. Emergency powers of the executive may be utilized to enforce them.

In addition, all decisions of councils are
subject to cancellation on grounds of illegality or
exceeding local powers by order of the Council of
State in an action initiated by the prefect or

Ministry of Interior. Orders and actions of mayors
and local staff are subject to cancellation by the
prefect if they conflict with national law or executive orders. Formal exercise of these legal powers
is seldom necessary.

Control over elected officials is another aspect of supervision, which in some cases has significant political implications. Mayors can be suspended for a month by the prefect, and occasionally
are, for failure to fulfill duties or for acting
contrary to the interests of good government and
orderly administration. The minister of interior
can suspend a mayor for up to three months or, in
serious instances, dismiss him. (If a strong political majority exists in the local council, however,
the higher authorities may have trouble on their
hands. In the postwar period when Communist mayors
of several communes in Seine were suspended for
three months, the assistant mayors refused to take
office and the minister had to appoint a temporary
administrator until a satisfactory compromise was
worked out.)

Councils, other than the council of Paris, can
be dissolved only by decree of the President of
France for a reason that is justifiable in the
administrative courts. While the prime minister
can dissolve the Paris City Council without explicit justification, the power is virtually unused.
Individual councilors can be unseated for specific
reasons, generally ineligibility or failure to perform duties.

Finally, as has been pointed out, the national
government has issued regulations governing administrative procedures and local personnel that are
enforced through administrative supervision. Moreover, the senior staffs of the city and Seine belong now to the national civil service, subject to
regulation from the prime minister's office.

Technical Control

Measures for central-government control over the substantive aspects of public services and works undertaken by local authorities are not codified or systematized, but they have grown in scope in the postwar period, as all units involved in urban government have become increasingly engaged in technically oriented activities. Standards are set by the technical ministries (e.g., industry, agriculture, public works and transport, construction, public health) and may be supplemented by additional local regulations. Inspection in terms of such standards is carried out by field services or inspectorates of the ministries. For example, the type, space, and amenities of schools and public housing built by local authorities are spelled out in detail by national law and regulations, enforced by national inspectors.

Technical control is exercised beyond the scope of standards when local agencies apply for project grants directly to the relevant ministry. The financial decision thus required engages the ministries, and sometimes the government or Parliament, in decisions respecting the desirability of the project itself in both its political and its technical aspects. Where and how Paris will get badly needed additional water supplies has been a subject of national-government debate for years. Without agreement of relevant national officials on this, adequate capital financing is not available to the city for major expansion works. Project proposals are generally revised by the national field services; indeed, those for smaller communes without extensive technical staff of their own are actually designed for them by a national service.

Moreover, the increase in the complexity of municipal problems and services accompanying urbanization has rendered local agencies informally dependent upon higher government technical aid and advice. The centralizing tendencies of the bureaucracy in France have been reinforced by this dependence.

Increasingly, local agencies look to ministerial field staffs for technical guidance and policy
alliances more than to the mayor and local political
officials who vote the budget they are operating
under. (A simultaneous development has been the
growth in the role and influence of the local
secretary-general, or chief administrative officer,
as the mayor's job becomes more complex.) An ex-
treme example is that of the Bridges and Roads services (Ponts et Chaussées), which have a near monopoly of trained personnel and equipment for highway
construction on which local roads agencies depend.
This bureaucratization of local activities is closely correlated with spreading urban development,
which has altered the tasks of municipal authorities
in the Paris Region and other urban areas throughout the world.

It is precisely this tendency to vertical
alliances that, together with the traditional independence and rivalries of ministries and their
bureaus, has undermined the effective coordinating
powers of the prefects and has raised problems of
coordinated regional action. Even when the prefect
succeeds in influencing a field-service chief, he
often is overridden by central-ministerial officials. These tendencies are far stronger in the
Paris Region than elsewhere in the nation.

Special Authorities

The role of centrally appointed officials in
urban administration in the Paris Region is expanding in conjunction with the use of special authorities. Designed to provide appropriate administrative frameworks for providing urban services, the
semiautonomous decision centers à fortiori cut
into the effective roles of locally elected officials. In addition, as authorized under French
law, most of the special authorities utilized
engage special forms of higher-government supervision, and many are dominated by higher-government

appointees. Intercommunal special districts and
local public-housing authorities, although created
by local initiative, for example, are closely regulated by national law and supervised by prefectoral
authorities. One fourth of the governing board, as
well as the director and treasurer of intercommunal
corporations (régies), are appointed by the prefect. At the regional level, central-government
control is more pronounced. In the Paris Region
centrally appointed officers dominate the regional
planning agency (the district), the regional transport agency, and the regional land-acquisition
agency.

THE BEHAVIOR OF THE INTERGOVERNMENTAL SYSTEM

While the assigned responsibilities and fiscal
resources of local authorities in the Paris Region
are commensurate with those of other French local
authorities, with some exceptions, higher-government
controls over them are more pervasive and more
varied. Such controls are stronger in the Paris
urban area than elsewhere, not only because of the
special structures for the city and Seine, but also
because of the nature of government activities in
the region. Administrative responses to urban problems
in recent years have tended to intensify central powers. This result flows from the national-
government control of capital finance and of intergovernmental institutions, as well as from the
superior technical capacities of the national civil
service. All of the major administrative changes in
the Paris Region of the past decade aimed at achieving coordinated development of the urban complex
and more efficient urban administration have increased the role of the national government. These
include the reorganization of départements, the
creation of the district, the reorganization of
transport regulation, the institution of urban
plans, the Interministerial Committee for the
Paris Region, and the creation of various special
authorities. The future role of the regional prefect and regional service authorities (should they

be created as authorized) will further shift the balance of power from locally elected officials to centrally designated bureaucrats. Whether any of these mechanisms will succeed in coordinating various bureaus and service hierarchies is a far more uncertain question.

At the same time, the flow of effective control is not entirely one-way. The political influence of local councils has been substantial at various times on various issues, and the successful implementation of regional plans is contingent upon their cooperation. Local powers are growing absolutely, if not relatively, as witnessed by the increase in borrowing powers of Paris and Seine and the slight easing of controls over their councils effected in 1961. Moreover, the regional prefect and work groups of the district, while structurally part of the national bureaucracy, have exerted pressure on the central government to heed the needs of the region and to receive more favorably the local requests for capital aid.

Informal mechanisms of intergovernmental relations are important, for in spite of the proliferation of formal controls, informal communication among the large number of authorities engaged in each decision process is a major mechanism in the operation of the governmental system in the region. A typical though simplified sequence of events for a major urban project is as follows:

1. Initiative is taken by a département, commune, or the City of Paris (the original suggestion often coming from a specialized operating agency); an architect or works director is appointed to draw up the proposal, and the council approves it.

2. The proposal is forwarded to the prefect, who may point out changes needed to conform to law or to meet application requirements.

3. The proposal file is sent to <u>all interest-</u>
<u>ed agencies</u>--ministry services, special authorities,
prefecture services, etc.--whose activities have
some bearing on the function involved. They all
attach their opinions. This is a standard procedure, which builds interagency communication into
administration but also slows the decision process
considerably.

4. If the project receives the support of the
ministerial bureaus involved, study and negotiation
on methods and feasibility of financing begin, covering possible participation of local authorities
through borrowing, national grants, district grants,
and loans from the Caisse des Dépots or other national loan funds (which are contingent upon agreement of the ministry or ministries involved to extend matching grant funds). The Ministry of Finance is crucially influential at this stage.

5. The agreed-upon financial participation is
formally authorized by each party, including passage of local investment budgets,which must be
approved. Continuing program authorizations must
be voted annually in the budgets until completion
of the project.

6. Implementation of the project is turned
over to an existing agency--local or departmental
service or special authority--or to a special mixed
corporation set up for the purpose, whose board of
directors includes representatives of the various
public agencies and governmental units involved.
The alternative is chosen in the course of negotiations and depends generally upon the nature of the
work and the initiating agency.

Omitted in this description is consultation
with special advisory groups that are established
for most important functions, such as public housing, education, planning, and transport. In an
average <u>département</u> in France, there are 125

committees that include representatives of the prefect and various service chiefs, as well as relevant
professional and interest groups. These are being
reduced in number in the Paris Region.

Negotiation, interagency bargaining, and overlapping membership of officials on study and advisory boards are central characteristics of this
process. Although national powers are crucial,
decision-making is not focused. The approval process involves a myriad of State officials and agencies, as well as local authorities. Strong leadership is necessary to provide the momentum to organize and carry through large-scale projects in such
a complex administrative system (in the past in
Paris this was provided by such forces of leadership as great kings and administrators, such as the
prefect Haussmann). This leadership and coordinating force was lacking in the Paris Region in the
postwar period while the government was underplay-
ing the needs of the Paris Region and concentrating
on development elsewhere. The attempt now is to
build coordinating and leadership vectors into the
administrative system through district planning,
the regional prefect, and the Interministerial
Committee.

The district activities have begun to stimulate action in the urban area by selective support
of proposals, persuasion, and dissemination of
information. Regional consciousness is developing.
In the district council, divisions of opinion are
more and more on technical questions rather than
along geographical and political lines, and emergent
public interest in regional problems is manifest in
the growing number of articles and conferences dealing with them. Since the creation of the district
in 1961, the per capita volume of public investment
in the Paris Region has quadrupled. This reflects,
of course, not only the district's activities but
also the shift in government policy that underlay
the district's creation.

Nevertheless, many local experts feel that the essential administrative problem in the area--that is, the requirement for simplification and focus of power to get things done--has not been mitigated. This problem lands one in the midst of the major issue of intergovernmental relationships in the Paris urban area: centralization.

The active debate concerning administrative organization in the Paris Region exists between the advocates of greater local powers and the proponents of reforms that increase the relative role of central authorities. The former are definitively the weaker.

The allocation of financial resources is at the heart of the controversy. Local politicians and officials object to the increasing burden of obligatory expenditure, the inadequacy and rigidity of local taxes, fiscal inequities among communes, and the complexities of obtaining grant and loan assistance through individual ministries and national-government funds. They complain that they can have no real independence without access to revenue sources autonomous from the national public-finance system.

Paris officials have long argued that Paris should have a mayor and administration of its own, as well as greater fiscal powers, maintaining that the city administration is controlled by national authorities unsympathetic to the needs of the city. This is fundamental to the objection of some to the 1964 administrative reorganization of the region. Although the legal changes separate Paris from the département, they maintain the government-appointed executives for the city administration. While the opponents of centralization hail the increases in the decision powers of the city council, and in the city borrowing powers, many feel that the fundamental dependence of the city on the State has not been affected. A member of the Paris council and head of

its budget committee (who also sat on the district
council) complained, for example, of the frustra-
tions of national control:

> The State which has finally authorized the
> city to borrow, has approved the budget by
> which the city decides to use part of the
> borrowed funds to build schools, and has
> controlled location decisions and techni-
> cal aspects of construction plans, refuses
> to pay its share in construction or is late
> in paying it.[2]

A suburban mayor and former president of the
Seine council pointed out in interviews for this
study that in relation to the growth of their tasks,
the local staffs in the region have expanded little.
He feels that the lack of understanding of the problems facing the Paris urban area on the part of
the supervising authorities accounts for a large
part of the lag in investment in urban facilities.

Finally, while most local and national officials acknowledge the needs for planning and greater coordination in the region, both among local
units and within the central government, many local
authorities object to the form of the district, particularly that its council is partially appointed.
Generally, however, local officials and politicians support measures for general coordination and
policy control, but object to detailed control and
supervision by higher authorities.

Notes to Chapter 3

1. Sources of fiscal data used are documents
published annually by the Ministère des Finances et
des Affaires Economiques, including Statistiques
des comptes des départements, des communes et des
établissements publics locaux; and Le Budget.

2. Alain Griotteray, <u>L'Etat contre Paris</u>
(Paris: Hachette, 1962).

CHAPTER **4** PLANNING AND PLAN
IMPLEMENTATION

In the past decade, regional planning of both
land and public-investment resources has become an
important aspect of the governmental effort to
adapt administration of Paris to urban-development
requirements. The planning undertaken to date has
comprised study of growth trends and needs for
urban public facilities, formulation of both urban
design and investment goals, and identification of
priority governmental programs. The plans for the
Paris Region correlate major service categories of
public activity and consider the long-range dimensions of policies. Hence, they attempt to counter-
act the fragmented administrative structure for
Paris and its inherent inertia. Resistance to
planning and coordination within the administration are persistent, but major planned projects are
under way and the new institutions for planning and
coordination are exerting pressures within the
governmental system that were notably lacking in
previous years.

Though a recent innovation, planning for the
region has already grown into a complicated process involving several government agencies and at
least four different kinds of plans (see Chart 4).
The District of the Paris Region is now the major
planning authority, but its activities are defined
by and contingent upon national planning policies.

Important from the point of view of the potential
for plan implementation is the fact that the structure of the district incorporates representatives
of most participants in administration: local
authorities, who are represented on its council;
the government, which is represented by the delegate general (henceforth regional prefect) and
which must approve district plans; and the bureaus
and services most intensively involved in the region's public services, whose personnel have been
both assigned to the district staff and represented
on district work groups.

FRENCH PLANNING SYSTEMS

There have been two major types of formal
planning in France during the postwar period:
macroeconomic and productivity planning, centered
in national four-year plans prepared by the General
Planning Commission (Commissariat Général du Plan,
or CGP); and aménagement du territoire, or town and
country planning,* involving long-term and local-
ized spatial projection of public facilities, land
use, and urban development.

The General Planning Commission is a staff
planning agency of the national government, directly responsible to the prime minister. It has considerable influence and intellectual autonomy but
no powers of plan approval or implementation. The
government selects the general objectives of four-
year plans in preparation (for example, economic
growth rates) and approves national and regional
plans in draft and final stages. Parliament votes

*This English translation of aménagement du
territoire has come to be accepted, but the French
term has wider meaning. Literally, it means management of territory. It encompasses all phases of
efforts to plan, control, and develop spatial relationships and physical facilities.

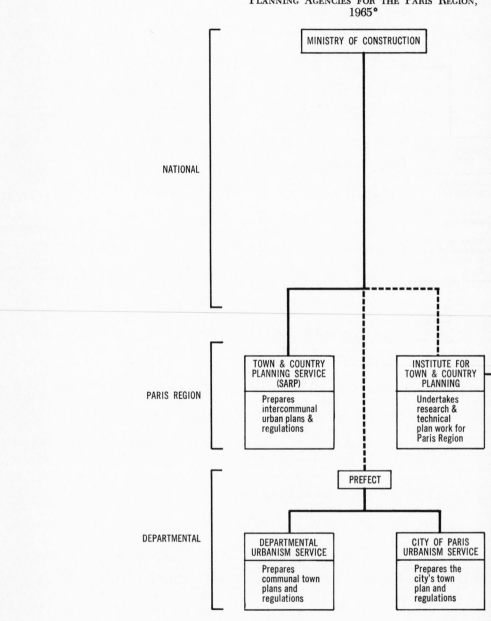

Chart 4
Planning Agencies for the Paris Region,
1965*

MINISTRY OF CONSTRUCTION

NATIONAL

PARIS REGION

TOWN & COUNTRY
PLANNING SERVICE
(SARP)

Prepares
intercommunal
urban plans &
regulations

INSTITUTE FOR
TOWN & COUNTRY
PLANNING

Undertakes
research &
technical
plan work for
Paris Region

DEPARTMENTAL

PREFECT

DEPARTMENTAL
URBANISM SERVICE

Prepares
communal town
plans and
regulations

CITY OF PARIS
URBANISM SERVICE

Prepares the
city's town
plan and
regulations

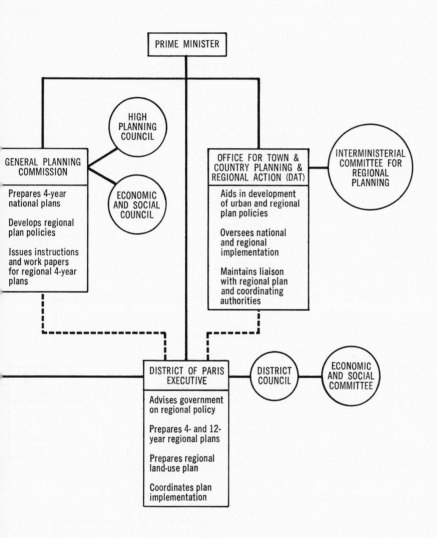

PRIME MINISTER

HIGH PLANNING COUNCIL

GENERAL PLANNING COMMISSION

Prepares 4-year national plans

Develops regional plan policies

Issues instructions and work papers for regional 4-year plans

ECONOMIC AND SOCIAL COUNCIL

OFFICE FOR TOWN & COUNTRY PLANNING & REGIONAL ACTION (DAT)

Aids in development of urban and regional plan policies

Oversees national and regional implementation

Maintains liaison with regional plan and coordinating authorities

INTERMINISTERIAL COMMITTEE FOR REGIONAL PLANNING

DISTRICT OF PARIS EXECUTIVE

Advises government on regional policy

Prepares 4- and 12-year regional plans

Prepares regional land-use plan

Coordinates plan implementation

DISTRICT COUNCIL

ECONOMIC AND SOCIAL COMMITTEE

*In 1966, the executive officer of the Paris District was made regional prefect, subject not to the prime minister but to the minister of interior. The Ministry of Construction was made a branch of the Ministry of Public Works.

approval of the national economic plan, which it
cannot modify. The planning commission utilizes
two major advisory bodies: the Economic and Social
Council, composed of private representatives of all
major socio-occupational categories; and the High
Planning Council, headed by the prime minister and
including representatives of labor and industry and
of regional development committees. In addition,
"modernization committees" with special functional
focus aid in the formation of segments of the plan.
They include both private and bureaucratic representatives. The four-year national plan outlines
both public and private investment policies and
macroeconomic targets.

The Ministry of Construction was responsible,
until 1963, for the development of town and country planning policies and the preparation of urban
plans. In fact, comprehensive regional and urban
policies were not developed prior to 1960, and the
only major national policy on urbanization was that
of decentralization, calling for the channeling of
urban growth and economic development to areas other
than Paris. The ministry supervised and participated in preparation of town plans in large urban
centers.

As the government, under the urging of the
General Planning Commission, began to turn its
attention to regional planning and coordination in
France, it recognized that the administrative sep-
arations between national and local planning,
between economic and physical planning, and between
planning per se and investment decision-making were
too great if the whole system were to achieve
rationally coordinated regional development. Most
investments contemplated by economic plans ultimately involve physical facilities on land.   The
intraregional spatial effects of these investments,
however, were not considered by economic plans,
and land-use controls generally were not based on
investment policies.   The French Government has

turned to regional project plans to bridge these
gaps.

The planning commission proposed institution
of regional project plans in 1955; in 1957, the
government promulgated decrees calling for their
establishment. The first efforts were halting;
the several regional plans prepared up to 1960
were too vague to be action programs and were not
well integrated with the national plan.

Since 1960, the government has instituted new
administrative arrangements for: (1) establishing
regional plans that will be operational transla-
tions (tranches opératoires) of the national plan,
both outlining general perspectives of economic
development in the twenty-one regions and identifying public investments to be made in them during
the four-year period; and (2) stimulating administrative coordination at national, regional, and
departmental levels for purposes of plan implementation.

The Paris Region has been subject to separate
decrees and laws for the most part, such as those
establishing the district and regional prefect and
reorganizing the départements. In each of the
other twenty regions, the new arrangements include:
an interdepartmental committee, comprising representatives of public interests and various technical services and operating agencies, to prepare the
regional project plans on the basis of documentation provided by the national planning authorities;
an ex officio regional prefect who is the prefect
of the département in which the administrative capital of the region is located and who has been given
certain coordinating responsibilities for purposes
of plan implementation; and a regional development
committee with an advisory role.[1]

Physical and economic planning are to be integrated both in the regional project plans thus

developed and at the national level. The major
administrative adjustment on the national level has
been the establishment, in 1963, of the Office for
Town and Country Planning and Regional Action
(Délégation à l'Aménagement du Territoire et à
l'Action Régionale, or DAT) in the prime minister's
office to stimulate and evaluate plan implementation at national and regional levels and to develop,
in close cooperation with the General Planning
Commission, long-term policies on urbanization and
spatial development that will serve as an integrating framework both for the national four-year plan
and for regional and local plans. The director of
the DAT, aided by a small staff, is to receive progress reports from ministries and regional committees, to maintain liaison with the coordinating
prefect of each region, and to report annually to
the government on national and regional plan implementation. He supervises and reports on regionalization of the national budget, which is to accompany annual fiscal regulations prepared by the
Ministry of Finance.

In summary, the DAT is a staff agency designed to maintain communication among regional and
national planning and plan-implementation authorities and to evaluate plan effects. It is to form
an articulating joint between regional and national planning and between economic and urban planning.

The Ministry of Construction, while it lost
general policy-development responsibility in town
and country planning to the General Planning Commission and the DAT, remained responsible for a
large part of the implementing activities in this
area and for preparation of local town plans. In
1965, however, this ministry was abolished, its
staff becoming a branch of the Ministry of Public
Works. This change further reduced the independence of urban land policy and housing from related
service activities and investment plans.

In theory, then, physical and economic

planning are to be undertaken and harmonized on
national and regional scales. While the institutions established in the Paris Region differ from
those in other regions, planning for it, like the
others, is to be implementive of comprehensive
national policies. This new thrust of government
effort has shifted attention from the negative
policy of national decentralization to that of
complementary regional development, which has
served as a far more constructive policy context
for administration in the Paris Region.

PLANNING FOR PARIS

There has always been planning in the Paris
Region, primarily by ministerial offices and operating services for particular sectors, such as education and manpower. What is new is planning for the
region: preparation of formal, multipurpose plans
with explicit regional coverage by a permanent comprehensive planning organization for the region.
Three kinds of regional plans have been prepared:
(1) a four-year investment plan for public projects (the tranche opératoire of the national economic plan); (2) a twelve-year urban-development plan
projecting growth and investment needs and outlining long-term development policies and projects;
and (3) a regional land-use plan setting forth
broad policies and goals of urban form. They over-
lap in subject matter, of course, but differ in
time range and specificity. In addition, communal
and intercommunal town plans throughout the entire
region have been required by law since 1919. Their
content is henceforth to be based on the regional
land-use plan as well as urbanism regulations set
forth by the construction branch.

Four- and Twelve-Year Plans

The four- and twelve-year plans are the responsibility of the staff of the district. The

first four-year plan (1962-65) was completed in
December, 1963, and the twelve-year plan was completed by 1965. The four-year-plan chapters were
prepared by thirteen functionally specific work
groups composed of administrators and technicians
from relevant national and local agencies (mainly
national). Such utilization of staff from operating
bureaus builds interagency participation into the
plan-development process. Virtually no significant
public involvement or interest-group pressure came
into play in preparation of this plan, although the
district has begun public-opinion polls and public-
relations campaigns. In each case, the magnitude of
investments planned and the projects included were
chosen within the general confines of the national
plan, government policies, and ministerial opinions.

After drafts of the plan were submitted to the
district council and to the prime minister's office,
suggestions from both were acted upon in revision.
In addition, the prefects of the three départements
participated in monthly progress meetings with the
delegate general to the district. The final ver-
sion was, in the words of the delegate general, "a
synthesis of the reports and opinions of all parties
concerned." It was submitted to the district council for final opinion and formally to the prime minister. The plan was subsequently approved by government decree.

Actually, most of the major projects included
in this plan had been proposed and elaborated by
various local and ministerial operating services
prior to the planning work, which was started late
in the plan period. The plan, however, served to
integrate various proposals and to establish priorities. It surveys the needs in the region in major
service categories and sets investment targets for
transit, highways, parking facilities, water and
sanitation, recreation, and cultural and educational facilities. Construction-rate targets are set
for housing, but investment in this sector is not
specified in the plan. Major projects are

identified for transit, highways, water, sanitation, recreation, and regional markets. Map 3 shows the major plan projects for the region. The targets for capital program authorizations in the period, 1962-65, are as follows:

	Million Francs
Mass transportation	2,709
Urban highways	4,908
Parking facilities	536
Navigation and ports	194
Energy	2,665
Seine reservoirs and flood control	196
Water supply	822
Sanitation	1,236
Public buildings and fire protection	418
Education (primary and secondary)	2,745
Public health	1,441
Recreation	572
Green spaces	239
Cultural facilities	497
Total	19,178

The dominance of transportation is striking; it accounts for over 43 per cent of the total.

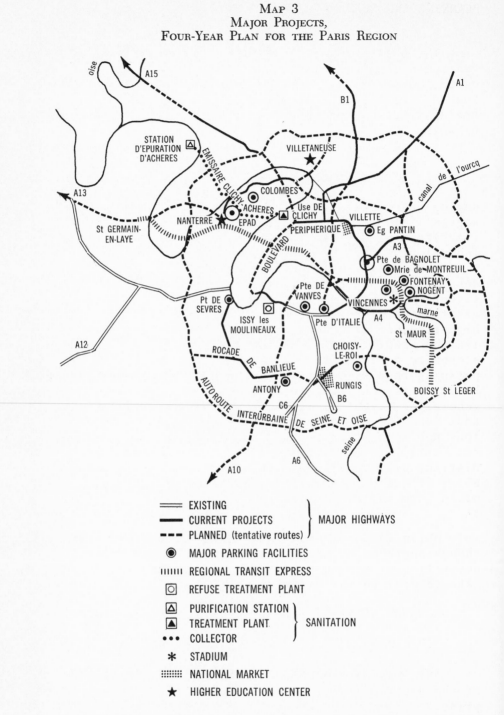

MAP 3
MAJOR PROJECTS,
FOUR-YEAR PLAN FOR THE PARIS REGION

EXISTING
CURRENT PROJECTS } MAJOR HIGHWAYS
PLANNED (tentative routes)

⦿ MAJOR PARKING FACILITIES

ıııııı REGIONAL TRANSIT EXPRESS

▢ REFUSE TREATMENT PLANT

△ PURIFICATION STATION }
▲ TREATMENT PLANT } SANITATION
••• COLLECTOR }

✳ STADIUM

::::::: NATIONAL MARKET

★ HIGHER EDUCATION CENTER

Source: Délégation Générale au District de la Région de Paris, *Programme Quadriennal d'Equipment pour la Région de Paris, 1962-1965* (Paris, 1963).

The twelve-year plan expands upon the four-
year plan in projecting longer-range growth forecasts, public-service needs, and investment targets.
Much of the basic research for it was conducted for
the district by the Institute for Planning and
Urbanism (IAURP).*

Regional Land-Use Plan

The first postwar land-use plan for the Paris
Region (PADOG) was prepared in 1958-60 by the Paris
Region field service for town and country planning
(SARP) of the Ministry of Construction (see Chart 4).
The stimulus for this effort came from the commissioner of construction and urbanism of the Seine
département, who became, in 1958, minister of construction. The central concept of the plan was the
creation of grands ensembles, which are the large
high-rise housing developments that now character-
ize the newer residential suburbs of Paris and
economize on land use. The plan was a vehicle for
pursuing this scheme for housing development in
Paris during the time the government supported a
general policy of limiting investment in Paris. In
addition, the plan summarized deficiencies in various public-service sectors and proposed some projects that have been expanded and incorporated into
district plans. Although it mapped land use generally, it did not formulate comprehensive goals of
urban form and development.

While PADOG did not have the force of law, even
though approved by government decree, it did orient
activities of the construction services and the delegate general to the district until 1963.

Responsibility for regional land-use planning

_____

*See Page 61 above. Note that the executive
of the district is also president of this enterprise.

has been transferred to the district staff,
which submitted the "Urban Land-Use Scheme for
the Paris Region" (Schéma Directeur d'Aménagement
de d'Urbanisme de la Région de Paris) to the district council and its advisory committee at a special meeting opened by the prime minister on June 22,
1965. (It is interesting to note that as soon as
news of the submission of the plan became known,
Prime Minister Pompidou was required to reassure
members of Parliament representing provincial areas
that it was not an indication that France would be
neglected for Paris and that similar plans would be
prepared for other regions.)

The plan, for which the Institute for Planning
and Urbanism did the bulk of technical work under
contract to the district, is premised on a regional
population by the year 2000 of 14 million (an increase of two-thirds over 1965 population of 8.5
million).* It does formulate general urban-design
goals. To overcome present space deficiencies in
the urban center (densities considered too high,
narrow streets, lack of recreation facilities,
scarcity of land for new facilities), the planners
propose to extend the urbanized portion of the region from approximately 460 to 850 square miles
while maintaining forests and selected open spaces
within those boundaries. New development is to be
concentrated along two east-west axes hugging the
rivers to the north and south of the present urban
complex. The most ambitious goal of the plan is to
create eight new urban centers of some 100,000 residents with employment and retail facilities, hospitals, higher-education institutions, and entertain-
ments in the outer suburbs, as well as to renovate
several centers in the inner suburbs in order to

---

*The forecast presupposes far more rapid growth
of other urban areas of France, reduction of net in-
migration into the region to nil, and continued
acceleration in the rate of natural increase. Present net in-migration is about 40,000 per year
(110,000 arrivals and 70,000 departures).

reduce the dependence of the region's population on
the central city. The plan lays out an extensive
highway and transit system of regional scope to
serve such a pattern of development.

The land-use scheme will serve as a framework
and guide for both local town plans and future
regional-project plans.

Town Plans

Communal or intercommunal plans have been
approved to date for every local jurisdiction in
the region save the City of Paris. Communal plans
in Seine were prepared by the département urbanism
service within the prefecture, and intercommunal
plans were prepared by a regional field office
(SARP) of the Ministry of Construction. Each local
plan is submitted to mayors and councils and several
public agencies before it is officially approved by
the departmental prefect and henceforth, the regional prefect. (That for the City of Paris has been
prepared by the city urbanism service in the prefecture and must be approved by interministerial
order.) The expenses of preparation are borne by
the national government, which made such plans
mandatory in 1919.

Generally, these plans and their supplementary maps and regulations are land-use proposals
and regulatory zoning and building documents that
implement construction regulations. They have not
entered into the realm of local capital programming.
They are now explicitly required to follow the proposed lines of the regional land-use plan, although
the latter does not yet have juridical status. The
approval process is depended upon to assure such
complementary relationships.

A Planning System?

Linkages among the plans for the Paris Region,
and between them and national plans and policies,
have been incorporated into planning procedures.

There is to be a means-ends hierarchy of plans that integrates physical and economic policies. Local town plans are to implement the regional land-use plan, which incorporates the aims of national policies formulated by the General Planning Commission and the DAT and detailed by the construction services. Regional four- and twelve-year development plans should translate the principles of urban design set forth in the land-use plan into investment needs and coordinated capital programs and at the same time serve as regional applications of national four-year plans.

The mechanisms for assuring articulation among plans include: the key involvement of the district executive and staff in each regional plan effort; interagency consultation and overlapping membership on planning staffs; and approvals by the prime minister (who has the benefit of advice from both the General Planning Commission and the DAT directly in his office). It is simply too early to tell, however, whether what in fact is operating is an integral planning system or a collection of planning operations. The timing of the various new planning efforts has not yet given opportunity for sequential formulation of plans from long-range urban design to short-range capital programs, but the cumulative and mutually reinforcing characteristics of the various plans prepared by the district are nevertheless evident.* The district four-year plan (its first and hurried effort) is essentially a catalog of major projects without unity of conception as to the future shape of the region. It is expected, however, that the new land-use plan and twelve-year plan will supply such unity of concepts

---

*The district has called upon and integrated the conclusions of a wide range of disciplines, including demography, economics, sociology, architecture, and public-service technology.

for future short-term investment programs.

It is not difficult to surmise, however, that
even with these mechanisms, harmony among plans will
not be automatic. Every planning effort with different scope and focus raises issues in a different
context. The scale of interests in the national
political arena differs from that in the regional
arena, particularly in light of the sensitivity of
the national government to accusations that the
Paris Region is favored and of protestations in the
Paris Region that it is ignored and bullied. The
district four-year plan points out, for example, that
the investment projections of the corresponding
national plan (Fourth Plan of France, 1962-65) for
the primary schools in the Paris Region are not adequate to meet demands. On the other hand, an advance
report on the fifth national plan proposes: (1) to
treat the Paris Region as a unique entity; (2) to
utilize a coordinated program of investments and
community-facilities development; and (3) to increase
the role of user charges in financing urban services.
Officials in Paris have pointed out that these policies are far too general to serve as an operational
framework for the next regional-investment plan.

Hence, in spite of the fact that higher-level
planning is supposed to control lower-level planning,
there is ample room for bargaining and adjusting the
framework at each stage. In a document of June, 1961,
dealing with the Paris Region, the General Planning
Commission had presented the framework for the district four-year plan, specifying the magnitude of public investment for various categories of public facilities. This framework was approved by the Interministerial Committee for the Paris Region. The district
four-year plan did not depart radically from it, but
adjusted almost every investment target upward,
corrected several estimates of demand (e.g., for
schools), and chided the government for failing to
meet, in 1961 and 1962, even the investment targets
set up by the planning commission document. The
thrust of the district four-year plan, which is to
point out that the investment lag in the region was
far greater than the national authorities had thus
far recognized, is testimony to the independence of

its viewpoint in spite of the fact that it was prepared by the staffs of national agencies under the
direction of a government appointee charged with
working within the targets set by the national
planning body. It is interesting to note that a
regional organization that is essentially a
national-government organ nevertheless rapidly began to articulate and represent interests of the
region more effectively than some regional organizations in other countries that are federations of
local authorities.*

In any case, the responsibility for consider-
ing utilization of fiscal and land resources, as
well as national and regional investment programs,
has been assigned to the two closely related plan
agencies of the national government; moreover, the
district is explicitly charged with dealing with all
kinds of planning for the region within the framework of national policies. These facts clearly increase possibilities for development of comprehensive and complementary goals and programs.

IMPLEMENTATION OF PLANNING CHOICES

How and with what effect do the products of
planning efforts enter into the administrative
system described in Chapters 2 and 3? The regional-
planning activities have been superimposed over
existing decision-making and operating structures of
government in the Paris Region, none of which have
been radically changed to adjust to them.

---

*The comprehensiveness of its approach to re-
gional development is also notable. Michel Piquard,
assistant to the district executive, has expressed
the primary goals of district planning as increasing variety and freedom of choice among styles of
living in the region while reducing fatigue and
frustration in urban life.

Furthermore, none of the plans except local town plans have the force of law over citizens or public officials, and the district itself at present has only the powers of persuasion and influence, aside from the limited latitude for action permitted by its budget.

Fundamentally, the capability to implement existing plans exists within the government structure. The district four-year plan goes to some lengths to examine the fiscal and administrative feasibility of its proposals and to spell out how each major project could be financed and by what units. The back-up technical work of project design is proficient. Finally, legal techniques for controlling development have been enacted. The question remains as to what extent those who vote and approve local budgets and commit national funds, on the one hand, and those who apply land-use controls, on the other, are willing to exercise these powers in such ways as to implement the formally approved regional plans.

There are two barriers to a fully affirmative answer to this question. The first is the resistance to automatic application of plan choices on the part of the many administrative agencies involved in government of Paris. There is no doubt that the degree to which decisions are determined by comprehensive planning choices is inversely related to the capability of the participants in urban administration to adjust decisions, both budgetary and bureaucratic, to contingencies of political pressures--particular agency goals--or new problems and circumstances.

Comparative analysis of the case areas studied suggests the hypothesis that, all other things being equal, the more fragmented the decision-making structure for the urban area and the more divorced the key decision-making authorities from plan preparation, the less likely are plan choices liable to be carried through the required series of implementing decisions. While plan preparation for Paris

involves most of the key authorities, decision-
making for the region remains highly fragmented.

Some of the measures being instituted in Paris
to facilitate implementation of plans are explicitly
aimed at reducing such fragmentation (or, conversely,
at concentrating power): the coordinating powers
of the new regional prefect, the interministerial
committees, and the DAT, for example. A rich subject of study over the next decade is the degree to
which these actually alter patterns of decision-
making in the region. Other measures establish
specific legal procedures that make certain decisions contingent upon plan items (such as approval
of certain construction permits by plan officials
and conformance of local-investment budgets to plan
priorities).

In any case, some instances of failure to
implement plan choices reflect not only resistance
to coordination by officials, but also legitimate
trade-off among several valid policy goals. This
is the second obstacle to automatic application of
plans. Such interests as fiscal stabilization that
are not considered in regional planning may be
brought to bear in the decision-making stage,
wherein plan choices are altered.

The primary governmental activities required
to implement the plans for the Paris Region are of
two types: (1) investment in and construction of
public projects; and (2) regulation and control of
land use and transfer.

Execution of Planned Projects

Whereas in the other regions of France one of
the departmental prefects of the region is "coordinating prefect" for purposes of stimulating regional
coordination for development-plan implementation,
the delegate general to the district has had that
role in Paris. His efforts to consult with prefects and to accommodate the district council,
which is composed of local councilmen, have

succeeded in reducing local hostilities to the district and in raising support for the four-year
investment plan. (At the same time, the projects
outlined in this first plan are so basic to the
most visible needs of the region and the lag in
investments was so great in prior years that the
potential for rejection by local authorities was
extremely low.)

Implementation of plan projects depends on the
capital and general budgets of the City of Paris,
the three départements, and to a minor extent other
communes and special agencies, on the one hand, and
allocation of national grants and investment budget
funds,on the other. In the district budget documents for 1964, the delegate general concluded that
the City of Paris had completed 83 per cent of its
plan commitments and the Departement of Seine, 86
per cent of them,to that time (or conversely, they
were 17 per cent and 14 per cent behind the plan
schedule respectively). The 1964 capital budget of
Seine includes 63 per cent of the commitments that
the district plan indicated for that département
for 1964.

Since 1961, separate investment sections have
been required in the budgets of Seine and the City
of Paris. These are comprised of four-year capital
programs with annual capital-budget breakdowns.
They are supposed to be explicit implementations of
national and district four-year plans. The 1964
capital budgets of these two units, however, did
not constitute operational applications of the district plan. They mainly included program authorizations for previously started projects, some of which,
however, had been mentioned in the plan (including
work on the peripheral highway by Paris and reservoir work by Seine). In the future, analysis of
these budgets in terms of plan implementation is
supposed to accompany their submission to the government for approval.

The operating budgets of the city and the three
départements, which are prepared by the prefects on

the basis of estimates presented by the operating
agencies, are not linked to any formal policies-
planning process.

When one recalls, however, that a major explana-
tion for the prior lag in public investments was the
failure of the national government to commit national funds to Paris area projects, it becomes clear
that the success of the district officials in mobil-
izing regional interests and advocating plan projects to the national government is crucial to plan
implementation. The four-year plan calls for capital expenditure of 19.2 billion francs in specified
programs. Of this amount, 11.8 billion francs were
allocated to governmental units according to budgetary source   (see Table 3). Some 37 per cent was
to be directly charged to the national budget. In
addition, of course, most of the local and district
participation is dependent on authorization of local
borrowing and budget approvals by the national government. The delegate general estimated in 1964
that 91 per cent of the planned national commitments to that time had been made. This represents
considerable improvement over the record of national-
government implementation of its own prior targets
for the region covering 1961 and 1962. Only about
60 per cent of national investments had actually
been authorized in those years.

The total public-investment record of the
nation in the region since 1960 represents a sizable increase over former years. The total per
capita volume of public investment in the region
has quadrupled since the creation of the district,
as has been noted. The district and its plans
cannot be identified as the major cause of this
trend, however, for the shift in government policy
toward the Paris Region preceded and underlay its
creation. In large part, the efforts of the President of France together with general economic expansion have augmented resources made available in the
region. Planning accompanied decisions to spend and
to develop. Nevertheless, efforts to implement district plans presently engage district authorities in

DISTRICT FOUR-YEAR PLAN
ALLOCATION OF FINANCING
(millions of francs)

Programs	Total	National Government	City of Paris	Départements		Communes and Their Special Districts	District of Paris	Others
				Seine	Other Two			
Public Facilities[a]								
In Urban Communes	1,902	125	270	583	32	594	183	115
In Rural Communes	192	27	-	-	7	97	44	17
Management of the Seine River	390	216	10	108	15	6	23	13
Public Buildings	375	153	62c	-	-	151	9	-
Fire Protection	43	16	5	-	-	18	3	-
Green Spaces	239	18	81	64	19	-	17	40
Sports	572	157	99	-	17	208	56	35
Cultural Facilities	497	497	-	-	-	-	-	-
Subtotal:	4,210	1,209	465 62c	755	90	1,074	335	220
Parking	536	-	-	350	-	-	-	186
Highways	4,908	2,588	1,197	548	78	60	414	23
Transit	2,112	553	-	-	-	-	553	1,006
Total:	11,766	4,350	1,662 62c	1,653	168	1,134	1,302	1,435
Education[b]	2,745							
Health[b]	1,441							
	15,952							

Source: Délégation Générale au District de la Région de Paris, Programme quadriennal d'équipement 1962-65.

a. Includes water supply, sewage, and public utilities.
b. Not definitively allocated as projects.
c. Shared item between City of Paris and Département of Seine.

continuing negotiation with the Ministry of Finance,
the Interministerial Committee, and other authorities to carry out and sustain these investment policies. That the government has committed itself formally to plan proposals and that the regional prefect presents the problems of the region to the
committee generate continuing pressures on the
decision-makers on behalf of plan projects. There
still occur instances in which the national government does not release funds in the amount agreed
upon, however. These are frequently the result of
decisions of the finance ministry, which justifies
its stand on grounds of stabilization policy and
national priorities.

The district four-year plan summarizes implementation problems as follows:

The realization of the four-year program demands an effort which, with the possible exception of urban transit, is compatible with
the growth of investment in public facilities
by the State proposed in the national four-
year plan and within the financial potential
of the Paris Region. This does not mean that
its proposals will be realized nor that the
financial resources will be mobilized without difficulty. The whole program brings to
light a certain number of obstacles which
only the modest level of policy goals in the
past permitted one to ignore....

In most categories, the program can be
realized under two conditions: The State
must accord, within the framework of investment credits corresponding to national plan
objectives, a part to the Paris Region which
is consonant with its national importance and
the deficiencies in public facilities which
are, on the average, greater here than in
other urban areas of France; and local authorities must support the part of the investments which falls to them and accept a

decisive role in implementation of urban
public works.[2]

The plan cites the conclusions that program
implementation requires that the proportion of
ministerial grants to local authorities for particular purposes, such as housing and highways, going
to units in the Paris Region would have to be increased; loan authorizations, particularly for
Paris and Seine, would have to be stepped up; and
general local finance should be improved. The
inequities among parts of the region as to local
revenues and local per capita tax rates were scored
as limiting the potential role of local authorities
in the areas of greatest investment needs.

Finally, the district itself is strategically
committing its own relatively small resources (200
million francs per year, plus borrowing as allowed
annually) in grants to communes, départements, and
special authorities to speed up priority projects
and stimulate action by other agencies.

The total program authorizations voted by the
district council in 1962-65 under the four-year
plan amounted to 1,360 million francs, of which 750
million francs were available for transfer in that
period (680 million francs from district tax revenues and 70 million francs from borrowing). The
authorizations voted apply to works and transfers
that continue into 1968.

About two thirds of the authorizations voted
in the period are allocated to major transportation infrastructure (in line with the thrust of the
plan as a whole), with priority given to mass-
transportation facilities.

The breakdown of the aggregate of the four
annual district budgets (1962-65) by functional
categories is as follows:

	Million Francs
Mass transportation	550
Highways	350
Parking facilities	65
Water, electricity, sanitation	162
Health and social facilities	17
Recreation and sports	23
Public buildings	5
Fire fighting	6
Flood control	19
Street and highway lighting	9
Green spaces	33
Research and planning activities	33
Land acquisition	24
Communal aid*	60
Total	1,356

These budgets were prepared by the district staff with the aid of the departmental prefects and service chiefs, voted by the district council, and approved by the ministers of interior and finance. The council brought about some changes in budget proposals, mainly reductions on controversial items, but its influence was generally exerted more in the plan-preparation stage than the budget-approval stage in spite of the fact that its formal power is limited to the latter.

As a result, district budgets have conformed fairly closely to the role contemplated for them in

---

*The last item was for general aid to selected communes--those fast-growing suburban municipalities with most severe financial problems-- to help them amortize loans from the national munici- pal loan bank (Caisse des Dépots). This item was not approved, however, by the supervising authorities (interior and finance ministries.)

the four-year plan.* In 1964, legislation passed by
Parliament reorganizing the départements in the
region included a provision that the ministers of
interior and finance could inscribe expenditures
for "priority projects" into the district budget if
the district council, after two successive delibera-
tions, had refused to vote such expenditures. If
local authorities do not come to agreement on carrying out such projects, the Council of State can order
that they be carried out by the district directly.
Limitations on the district's capacity to carry out
its part of the four-year program, however, have
come not from the council but from the national
government in the form of restrictions on its borrowing. Moreover, utilization of the credits voted by
the district has been slowed because of project delays in other agencies, particularly delays in land
acquisition and release of funds from the municipal
loan bank.

A key aspect of plan implementation to date has
been the delegate general's participation in various
decision-making processes. He has approval powers
over the regional land-agency projects, local urban
plans, and certain construction permits, for example. Prior to 1966, when his post was transformed
into that of regional prefect responsible to the
minister of interior, he was a staff aid to the
prime minister. He tends to dominate discussions of
the Interministerial Committee for the Paris Region,
in which he contributes the greatest knowledge and
information respecting the region. Following the
general French practice of interagency consultation,
the region must seek his opinion in the course of
many other administrative decisions. Finally, as
regional prefect, he has been given certain

_____

*Comparison of these budget figures with the
expenditures allocated to the district in the plan
(see Table 3) indicates that some shifts were made
between categories, but, in general, the budgets
conform to the plan.

formal coordinating powers over prefects and service chiefs (see Chapter 2).

Ultimately, however, the informal efforts of
the district officials to persuade local and central units to cooperate in plan implementation and
to mobilize support in the region are crucial. For
on balance, plan implementation depends on the
exercise of financial decisions by the City of
Paris, the départements, the ministries, a few special agencies, and above all the Ministry of Finance. At the operating stage, only one new agency
has been created for undertaking plan projects--and
that is to embark upon land acquisition. Thus far,
the regional transport agency, the departmental and
municipal services (such as highways services), and
the intercommunal water districts are the major
operating agencies for plan projects.

The district itself is breaking logjams of
intergovernmental financial negotiation both at the
stage of drawing up the plans and at the stage of
authorizations and approvals. Evaluation of the
effect of the total planning system, however, must
await the passage of time and particularly the impact of the next four-year plan (1966-69), which
will be keyed into the fifth national plan and the
regional land-use plan. The first four-year plan
was not completed until halfway through the period
it covered, and the record of budgetary fulfill-
ment is therefore not very enlightening; some of
the projects were actually approved or started before they were included in the plan.

The regional land-use plan--and particularly
the new urban centers it calls for--posits far larger
plan implementation tasks. At present, the district
is financing studies by the Institute for Planning
and Urbanism (IAURP) on feasibility and design for
these undertakings, which, according to current
thinking, will be assigned to special development
corporations.

Control of Land Use

The effect of planning for the region on major

public-investment programs has been examined, but
there is an important second aspect to plan implementation in the Paris Region: control of private
and semipublic development and general land use.
This is particularly relevant to implementation of
the twelve-year and regional land-use plans, but it
also affects the short-term-investment program, in
that cost is a major factor in feasibility of
investment schedules, and a large factor of project
cost in the Paris Region is the price of land. Moreover, requirements for community facilities are in
the long run determined by the rate and spatial distribution within the region of residential and
employment growth.

Land control has been exercised by the Ministry
of Construction and by the financial and tax policies of the government. The government policy of
decentralization, despite recent modifications, continues to underlie regional planning in the national
context. Its application has consisted in putting
limitations on construction in the Paris Region and
stimulating public and private investment in other
regions. Special approval from the minister of construction is required for the issuance of permits in
the Paris urban area for construction of more than
500 square meters of office, industrial, or commercial floor space, or for expansion into space of
the same or greater size. Approval is to be based
on judgment as to the feasibility of the activity's
being carried on outside the Paris Region without
economic hardship. Supplementary measures enacted
in 1960 levy a fee on offices and factories given
qualified approval for construction in the region,
the amount of the fee varying with the zone within
the region where the construction or expansion takes
place. Positive incentives to decentralization have
included subsidies, tax concessions, and loan priorities to enterprises locating in or moving to areas
of reduced employment in other parts of the country.

While these policies have stimulated rapid
growth (both industrial and residential) of provincial centers in France, many people interested in

local government and urban planning are skeptical about their effect in limiting the growth of Paris. The economic grounds for location in the region have been unarguable and were accepted by the ministry in most cases where application was made. Nevertheless, the region might have grown somewhat faster in the past fifteen years without these programs. From their inception to 1963, some 1,279 "decentralization" projects have been aided with a corresponding employment of 230,000.* Including families and persons attached to other attracted activities, 800,000 to 900,000 people who might otherwise have contributed to the growth of Paris have been involved. The general trend of growth in Paris and shrinking of rural areas continues, but the faster growth of provincial urban areas must be added to the picture.

In any case, the special construction-permit procedure could be used to influence development patterns within the region in directions indicated by the general regional plan. To date, this has not been among its purposes.

In addition, the municipal town plans and corresponding zoning regulations that are supposed to be implementary of the regional land-use plan have the force of law and are enforced by the regular construction- and subdivision-permit procedure that also enforces the urbanism code issued by the Ministry of Construction and local building codes. All parties desiring to construct a building or undertake important modifications of one must submit to municipal authorities a request for a construction permit, accompanied by detailed building plans and locational information. This request is

---

*For the most part, these have not involved moving facilities from the Paris Region to elsewhere, but merely establishing new industry in the provinces.

submitted for examination to all interested agencies
(construction, urbanism, and planning agencies; sites
committees; and relevant public service agencies).
Subsequently, on advice of the departmental director
of construction, the mayor issues the permit. The
mayor cannot refuse to issue it if the advice is
positive and all legal conditions have been met,
including conformance to local plans. Within the
City of Paris, the director of the urbanism service
issues the permit. Any large-scale construction
(e.g., that entailing 500 or more dwellings) that
does not require approval of the ministry must be
approved by the executive of the Paris District.
Subdivisions require special approval of the prefect (the subdivider must provide internal streets,
lighting, and water-distribution facilities and
contribute to provision of other required public
facilities, sometimes by the ceding of land).

These procedures have serious drawbacks: They
are time consuming, taking several months and sometimes years, and they are complex, often involving
fifteen or twenty agencies. On the other hand,
planning authorities are given the opportunity to
evaluate development requests. Moreover, French
administrators believe that the involvement of so
many officials and agencies assures honesty in
application. Proposals for simplification are
being considered by the ministry, particularly the
creation of a construction-permit committee including officials from the interested agencies in order
to maintain the multifunctional approach but concentrate the approval procedure.

General control of land is nevertheless a
severe problem in the Paris Region. The French
Constitution establishes an absolute right of
private-land ownership, to which the people are
deeply attached. The government does not have
significant land reserves, and expropriation procedures are lengthy, with compensation as set by the
courts generally very high. Rapidly expanding demand relative to suitable supply of land for urban
development in the center of the region has

produced considerable land speculation and inflation in the postwar years. An arsenal of legal tools
to control land prices and recapture added values
exists: special taxes on the sale of land, fees
based on the costs of public facilities needed to
service private development, price controls, and
restrictions on short-term resale of land. The
central government, however, has been reluctant to
provide for rigid enforcement of these measures,
primarily because of the virulence of opposition to
them.

In recent years, some mechanisms for public
acquisition, control, and development of land have
been developed. Departmental land services have
been established to undertake expropriation proceedings for local agencies. Strong control over all
public-land operations is maintained by the Central
Control Commission for Land Operations in the prime
minister's office and by its subordinate departmental commissions (in all départements save Seine,
where the central commission has operated directly).
All local public-land transactions require approval.

The regional land agency (Agence Foncière) for
the Paris area was created in 1962 by ministerial
decree and was empowered to acquire land in advance for urban-development projects and to carry
out development projects for national or local
agencies. It is a public corporation responsible
to the ministers of finance and construction,
managed by a board that includes national and
locally appointed members. Not until 1966 did it
begin operating, however. The City of Paris, the
three départements, and other major communes also
utilize mixed corporations to carry out land
development and urban-renewal projects.

Finally, a new procedure for reserving land
for plan implementation has been authorized by
national law since 1959. Certain zones--priority
urban-development zones (ZUP's) and zones for de-
ferred planning (ZAD's)--are established by government decree. Thereafter, for a period of from four

to eight years, the national government and local
authorities have rights of pre-emption in the zones;
they can exercise a priority option to buy any land
therein. If they exercise the right, the price is
set either by negotiation or by judicial determination based on comparable prices in the year preceding the establishment of the zone. Development of
the zones can then be undertaken through either the
land agency or special mixed corporations with
approval of the regional prefect or national construction authorities. The national Economic and
Social Development Fund must be consulted on these
projects. Private construction projects involving
more than 100 dwellings can be required to locate
within ZUP's as a condition of permit approval.

This is a method for short-term freezing of
land prices and expediting land-use plan implementation. There are already some 1,000 ZUP's in
France. The 1960 regional land-use plan for Paris
outlined about twenty of them, of which some fifteen have been established.

In summary, then, considerable powers and tools
exist for plan implementation in terms of land
development. The centralization of control in this
area, with key roles of the construction service
and the executive of the district, affords the
opportunity for coordinated use of these tools. To
date, the major limitations have been, first, that
the rate of public-land development projects in
the region has been low due primarily to costs and
local fiscal weakness; second, that controls over
land prices and speculation have not been applied
consistently; and third, that prior to 1965 no plans
set forth comprehensive concepts of the future design of the region that could serve as a basis for
exercise of the myriad of administrative decisions
in such a fashion as to achieve explicit and comprehensive land-development goals. This last point
demonstrates the fact that plan implementation in
many cases is contingent upon the plan in question
being operational--that is, of such a nature as to
serve as a realistic guide to diverse decision-

makers. The new regional land-use plan represents
a large step forward in this respect.

Notes to Chapter 4

1. For analysis of these arrangements for
regional planning and organization see Institut
d'Etudes Politiques de l'Université de Grenoble,
Administration traditionnelle et planification
régionale (Paris: Librarie Armand Colin, 1964).

2. Délégation Général au District de la
Région de Paris, Programme quadriennal d'équipement
pour la région de Paris, 1962-65 (Paris, 1963),
p. 86.

3. See Jean Ribat, "La Decentralisation
industrielle," Urbanisme, No. 80, 1963.

CHAPTER **5** SELECTED URBAN
SERVICES

A brief examination of the manner in which
several urban services are provided illustrates the
roles and relationships of the institutional participants in government of the Paris Region. The
services selected cover a broad range of administrative arrangements in the Paris area. Water
supply is considered a local--mainly communal--
service. Intra-urban mass transportation has undergone transformation from predominantly local organization to predominantly regional organization.
Public housing is provided through semiautonomous
municipal and departmental authorities established
to carry out particular programs that are fully
spelled out in national law and subsidized by earmarked State funds. Finally, education is a national service for which local governments have some
obligatory implementing functions and concurrent
powers.

WATER SUPPLY

The major operating water authorities in the
Paris Region are the City of Paris Technical Directorate of Water and Sanitation, which has been
situated in the Seine prefecture; two intercommunal special districts in the suburban areas of the
urban complex; and various small municipal services.

The water-supply system for the central city,

including pumping stations, wells, reservoirs, aqueducts, and other facilities placed throughout the urban area, is owned and operated by the city as a monopoly public utility, under the budget of the city council. The city water directorate is headed by a member of the technical corps of the civil service and is subject, with other directorates, to the supervision of one of the secretaries-general of Seine, who is responsible to the prefect. The directorate is composed of six major services : canals and diversions (with five subdistrict offices); pumping (with five subdistrict ofrices); distribution (with four subdistrict offices); sewage and drainage; water purification; and major treatment works. Personnel are recruited by the prefecture according to national regulations and the senior administrative class now belongs to the national civil service.

This directorate has also functioned as a departmental service, under the Seine budget, primarily to undertake some operations on behalf of suburban communes (with commune contributions). When the Département of Seine is dismantled, however, under terms of the 1964 reorganization law, this directorate will be situated in the new City of Paris administration headed by a prefect of Paris and thus will be separated from administration of the new départements.

The special water-supply districts in the region are public authorities created by agreement of the communal councils involved, with ministerial approval. Le Syndicat des Communes de la Banlieue de Paris pour les Eaux (Syndicat de la Banlieue) is an association of 141 communes (68 in Seine, 69 in Seine-et-Oise, and 4 in Seine-et-Marne) covering a population of over 2.5 million. Le Syndicat des Communes de la Presqu'île de Gennevilliers includes 9 communes of Seine. Both are supported by municipal funds, and their powers are limited to those specified in their charters--to provide water services (supply, treatment, and bulk distribution).

Directed by boards of local council representatives
(two from each commune, irrespective of population),
they are subject to the supervisory authority of the
prefects; for example, the prefects must approve the
director chosen by each board. The boards' budgets
are subject to the same regulations as communal budgets, and their employees belong to local cadres.
They are regulated by a complex of statutes and decrees supplemented by administrative case law governing such matters as their contracts with private
corporations and financial transactions. Finally,
the major portion of the capital resources of the
special districts flows from national, departmental
and district grants for specific projects.

Most other communal services, such as those of
St.-Maur, Coulommiers, Maisons Lafitte, etc., are
monopoly public services established by the communes
themselves (en régie). Some other communes supply
water through contracts with the major water companies. The departmental services (in Seine, the
technical directorate of water and sanitation, the
dams and reservoirs service, and the bridges and
roads service) provide technical aid and undertake
major works for communes. These are financed by
communal contributions and national and departmental
grants. Finally, an exception because of the historical significance of the area and its fountains,
Versailles and its environs are supplied directly
by a national service: Service des Eaux et Fontaines de Versailles-Marly et St.-Cloud.

Outside the City of Paris, distribution of
water to the consumer is generally handled separa-
tely from bulk supply through local contracts with
the water companies. These companies include La
Compagnie Générale des Eaux and La Société Lyonnaise des Eaux et de l'Eclairage. The former is a
nationalized company that serves about ninety of
the communes of Seine, and the latter is a private
corporation. These are under contract to each commune and special district that they serve. They
also execute some capital projects under contract.

There is general agreement among water technicians in Paris that small urban and rural communes
in the region cannot provide water-supply services
efficiently under conditions of rapid growth, both
because of local fiscal and staff weaknesses and
because of technical problems. The water tables
and resources for major expansion of water supply
transcend the communal boundaries. Thus far, the
use of special districts and departmental services
has been the response to this situation, but the
outer communes of the region are relatively ill-
serviced and have been reluctant to join special
districts.

Financing difficulties of all units have been
aggravated by the failure to raise water rates to
cover operating costs, requiring that the water budgets be balanced from general local revenues. The
rates in each case are set by the relevant council
of the commune or intercommunal district. Those of
the Syndicat de Banlieue are .735 francs per cubic
meter, and of the City of Paris, .41 francs per
cubic meter (about 6 and 3 cents per 100 gallons,
respectively). In the 1964 budget, the receipts of
the Paris water service were 96,043,000 francs, of
which 89,072,000 were from the sale of water.

Appraisal of Water Supply

The water produced in the Paris Region meets
high health standards set by the Ministry of Public
Health and is immediately accessible to consumers
throughout the urbanized portion of the region.
About 82 per cent of all dwellings in the urban
area have potable water outlets within them, and
almost all others are near outlets. There is
strong evidence, however, that total effective demand for water is limited by over-all supply.

Peak production figures for the Paris agglomération in 1957 (for a population that was then about
6.7 million) of potable water were about 90 gallons

per capita:

City of Paris, spring and reservoir water	350,000 m^3/day
City of Paris, treatment plants	800,000 m^3/day
Syndicat des Communes de la Banlieue	750,000 m^3/day[a]
Syndicat de la Presqu'île de Gennevilliers	170,000 m^3/day[a]
Société Lyonnaise des Eaux	158,000 m^3/day[a]
Service des Eaux de Versailles	70,000 m^3/day[a]
Total	2,298,000 m^3/day

Source: Plan d'Aménagement et d'Organisation Générale de la Region Parisienne, 1960, Annexe "N."

a. 95 per cent of these consist of water pumped from rivers and treated.

In addition, about 450,000 cubic meters per day of untreated water were supplied by the City of Paris to industry.

This capacity was estimated to be short of 1957 potential demand for potable water by 400,000 cubic meters per day. Conservation measures during dry periods and risks of dry taps in parts of the urban area are common. It is generally held that if the departures of residents from the city in August were not as numerous as they are, shortages would become obvious.

The over-all growth of the region partially accounts for shortages in suburban areas, but rising consumption habits are contributing to them in the city and throughout. Presently, 66 per cent of the region's dwellings have neither shower nor bath, and 45 per cent have no internal toilet, but these

ratios have been dropping as virtually all new
dwellings have these facilities. Average per capita water consumption remained steady in Paris
from 1926 to 1954 but rose rapidly in the suburban
areas of intense new construction.

Because of the backlog of expansion needs,
realistic targets for increasing supply are conservative in terms of potential demand. The targets set by the district four-year plan as compared
with 1959 capacity are as follows (cubic meters
per day):

	1959	1965	1975
Paris	1,100,000	1,250,000	1,550,000
Outside Paris	1,350,000	1,900,000	2,600,000

The major problem in water supply in the Paris
Region, then, is the failure to make timely capital
extensions in the postwar period, which is attributable to fiscal weakness of the communes, limited
borrowing powers for the city, and the policies and
priorities of the national government relating to
allocation of resources to the region.

The City of Paris, for example, has been aware
of the need for additional supply facilities for
over thirty years. The system built by Haussmann
and Belgrand in the nineteenth century taps all
major springs within 100 miles of the city. In
1926, this was the major source of potable water
for the city. By 1954, almost 60 per cent of the
total supply was filtered river water with a high
chlorine content. Since 1930, the city council and
various prefects have advocated a project in the
Vals de Loire that would bring water to Paris from
the Cosne region. Government approval and financing were never obtained because of investment policies and parliamentary opposition to having the
city encroach on these water sources. Several other
projects for the city have been proposed to the
government, but approvals were not forthcoming. In
1958, the government finally approved use of the

Vals de Loire waters for Paris but remitted the decision to further study three months later. The
city initiated studies of alternatives, contracting
with the Compagnie Générale des Eaux for research
and survey. Concurrently, the government studied
the possibilities of a purification station at Orly,
but the Ministry of Public Health raised objections
to the quality of water pumped from the Seine. In
the meantime, one of the alternatives being considered was jeopardized by the purchase of privately held
land in the Montreau region by an oil refinery, the
operation of which may ruin the underground water
supply in that area, according to the director of
the city and département water service. This alternative was subsequently dropped.

     This situation demonstrates the dependence of
the local authorities on the national government for
project approval and financial participation when-
ever major capital works are involved. The Paris
District concluded:

> ...with respect to water supply the
> resources of the Syndicat Intercommunal
> de la Banlieue are not adequate to per-
> mit it to undertake the works necessary
> to fulfill the rapidly growing demand
> of the 141 communes...of which the popu-
> lation has grown by 760,000 in eight
> years.[1]

     It is at the points of intergovernmental and
interministerial debate that the district is presently involved in trying to bring about agreement
on projects and break logjams in government decision-making. The Orly purification station is
included in the district four-year plan.

Planning and Investment Programs
for Water Supply

     Prior to 1960, planning for water supply in the
Paris Region consisted of project proposals prepared
within the various operating services. In 1960, the

TABLE 4

DISTRICT FOUR-YEAR PLAN
PROGRAM AUTHORIZATIONS FOR CAPITAL INVESTMENT
REGION OF PARIS, 1962-65
(millions of francs)

	Total	National Government	District	Départements	Communes and Communal Special Districts
Water Supply	821.67	40.70[a]	48.40[b]	4.80	727.77[c]
Seine Basin Control	196.16[d]	78.58	15.97	93.98	5.83
Sanitation & Sewage	1060.70	100.65	147.40	535.50	251.15

a. Includes 16.8 million by Eaux et Fontaines de Versailles; 8 million national participation in operations by the Syndicat Intercommunal de la Banlieue; 15.9 million participation by the Ministry of Agriculture in equipment of rural areas.

b. Includes 2 million for research and survey; 36.4 million for equipment of rural areas; 6 million for 20 per cent participation in construction of a filtration station by the Syndicat intercommunal de la Banlieue; and 4 million in grant to that syndicate.

c. Of which some 256.47 million by the City of Paris, to be financed by borrowing, with the costs of borrowing to be borne by the general city budget.

d. Includes the dam-reservoirs of Seine and Marne.

130

general land-use plan for the region of the Ministry
of Construction reviewed water resources and needs
and estimated over-all requirements for the subsequent fifteen years. Presently, district responsibilities with respect to water supply include sur-
veying needs and existing facilities, forecasting
future demand in the region, developing operational
investment programs and feasible financing agreements, coordinating the major works of various agencies, and aiding priority projects with district
grants. These activities are to be set in the framework of national water-resources policy--a sector of
town and country planning for which there is a special secretariat attached to the Office for Town
and Country Planning (DAT) in the prime minister's
office.

In the meantime, the effective water-supply plan
for the region is the district four-year plan, which
presents an investment program totaling 1,018.7
million francs on water-supply facilities and 196.2
million francs on Seine Basin control, the latter
including major reservoir facilities. The two categories account for about 5 per cent of the planned
outlay for 1962-65.

Most of the projects included in the plan
originated as proposals within various operating
water agencies. The water section of the plan was
prepared by the water subcommittee of the Public
Works Group of the district staff. This subcommittee is composed of personnel, mainly engineers, from
the City of Paris water directorate, the Seine
bridges and roads service, the General Planning
Commission, the Directorate of Local Authorities in
the Ministry of Interior, the Town and Country Planning Service for the Paris Region (SARP) of the
Ministry of Construction, and the Société Lyonnaise
des Eaux. The Seine Basin control section of the
plan was prepared by another subcommittee of the
same group. Neither included members from the
communes or intercommunal special districts. These
sections were submitted to the district council,
which reduced a proposed subsidy by the district to

TABLE 5

FOUR-YEAR PROGRAM FOR WATER SUPPLY
(million francs)

Project	Operating Authority	Capital Program Authorizations					
		Total	1962	1963	1964	1965	Thereafter
PRODUCTION							
Construct filtration station, Orly; modernize plant at St.-Maur; catchment facilities in Avre and Eure valleys	City of Paris	191.0	103.0	63.9	6.0	8.1	10.0
Reinforce filtration stations, Choisy-le-Roi, Mery-sur-Oise, Neuilly-sur-Marne; improve various plants (all will increase capacity)	Syndicat inter-communal de la Banlieue de Paris pour les Eaux (municipal special district)	179.0	31.0	1.0	51.0	17.0	79.0
Catchments in valley of Oise and improve various plants	Société Lyonnaise des Eaux & Syndicat de la Banlieue (company)	108.0	6.5	6.5	8.5	8.5	78.0
Extend treatment plant of St.-Maur	Town of St.-Maur	2.5	-	2.5	-	-	-
STORAGE							
Reservoirs of Lilas & de l'Hay-les-Roses	City of Paris	33.8	-	30.0	3.8	-	-

Increase capacity of various reservoirs	Syndicat inter-communal de la Banlieue (municipal special district)	12.0	3.0	3.0	3.0	3.0	-
TRANSPORT AND DISTRIBUTION							
Improve Paris network	City of Paris	41.67	12.97	10.7	4.2	13.8	-
Feeders of Marne, Seine, Oise, and various canal operations	Syndicat inter-communal de la Banlieue	240.0	30.0	71.0	27.0	82.0	30.0
	Companie des Eaux de Banlieue & Lyonnaise des Eaux (company)						
Network improvement		55.0	16.0	15.0	12.5	11.5	-
Network improvement	Eaux et Fontaines de Versailles and various communes	16.8	3.3	4.5	4.0	5.0	-
RESEARCH AND SURVEY		2.0	-	-	1.0	1.0	-
RURAL EQUIPMENT		136.9	45.0	37.9	28.0	26.0	-
Total		1,018.7	250.8	246.0	149.0	175.9	197.0

Source: Délégation Générale au District de la Région de Paris, Programme quadriennal d'équipement pour la région de Paris 1962-1965 (Paris, 1963).

133

the Syndicat de la Banlieue from 36 to 10 million
francs. Tables 5 and 6 summarize the programs of
both sections in the plan as published. They show
the authorities responsible for executing the projects, not those responsible for financing them.

These programs call for financial participation
from all units of government as shown in Table 5.
The total for water supply is less than the total
proposed investment--1,018.7 million francs--because
not all proposed expenditure had been allocated
among the participants.

While the local authorities are to provide for
almost 90 per cent of the water-supply investment
program, the major part of the 728 million francs
to be committed by them will be derived from loans
from national banks authorized by the government and
from regular capital aid from the ministries, mainly
the Ministry of Interior in the case of water. Direct national and département financing provides for
the major part of dam and reservoir construction that
comes under the heading of Seine Basin control.

As can be seen in Tables 5 and 6, the implementing authorities for the planned investment projects are the local authorities and contracting
companies with the sole exception of large-scale
flood-and-pollution-control works by the national
services. The most complex undertakings are to be
carried out by the Seine and City of Paris services
and the Société Lyonnaise des Eaux.

Negotiation for Plan Implementation

Although the initial agreements on projects,
priorities, and financing arrangements were worked
out among the various government units in the planning stage and received the approval of the government in the plan, the process of authorizations by
each unit involved is not without roadblocks. The
plan has no legal status and the district itself
plays a continuing role in persuading local and
ministerial authorities to take the implementing

TABLE 6

FOUR-YEAR PROGRAM FOR SEINE BASIN CONTROL
(million francs)

Project	Operating Authority	Schedule of Capital Commitments				
		Total	1962	1963	1964	1965
Local flood control works						
In Seine	Seine Department	20.0	5.0	5.0	5.0	5.0
In Seine-et-Oise	Communes and their special districts	15.3	.3	5.0	5.0	5.0
In Seine-et-Marne	Communes and their special districts	3.4	-	3.4	-	-
Bougival Dam	Communal special district	10.0	1.8	8.2	-	-
Work on La Basse-Seine	National services	15.0	-	1.0	7.0	7.0
Seine Dam-Reservoir[a]	Seine Department	-[a]	-	-	-	19.8
Marne Dam-Reservoir	Seine Department	118.9	3.0	17.0	98.9	-
Pollution control projects	National services	13.6	1.4	1.4	5.4	5.4
Total		196.2	11.5	41.0	121.3	22.4

Source: Délégation Générale au District de la Région de Paris, Programme quadriennal d'équipement pour la région de Paris, 1962-1965 (Paris, 1963).

a. Other capital commitments for the Seine Dam-Reservoir had already been made.

135

decisions.

For example, one of the most important new projects proposed by the Public Works Group in the four-
year district program is construction of the Marne
dam and reservoir, entailing a total investment of
118.9 million francs--54 million from the State and
64.9 from the Département of Seine. As of 1963, the
planned investment schedule was 3.1 million francs
in 1962, 5 million in 1963, 9 million in 1964, and
82.9 million thereafter, with a target date for system operation of 1970.* The project will reduce the
level of flood tides and increase reservoir and pumped-
water supply. The Paris Region will receive the
benefits of both of these effects, and the flood-
control aspects will also benefit parts of the
Champagne Region. The project is essential to ful-
fillment of regional water-supply targets for 1975.

The history of this project is a complicated
one. In past years, the Département of Seine has
constructed (with 45 per cent grants of the
national government) similar facilities of lesser
scope, leaving a good deal to be done to attain
meaningful goals of flood control and water supply.
The département decided to embark on a larger program involving the Seine reservoir and the Marne
reservoir, for which it drew up detailed proposals
in 1958-59. The construction of the former was declared to be for a valid public purpose for land
expropriation by the Council of State in September,
1959 (after an extremely dry summer); the expropriation order was issued in June, 1960; the work is in
progress and will be completed about three years
behind the original schedule. On the other hand,
administrative procedures on the Marne reservoir

----

*This schedule is a revision of that published
in the four-year plan and shown in Table 6, which
contemplated earlier and more rapid commitment of
funds.

project were blocked by the government in response
to opposition of local authorities in the area of
the construction. Finally, however, the proposal
was accepted on June 5, 1962, by decision of the
minister of public works, who then requested the
proper service within the ministry to prepare the
papers for the investigation into the aspect of public purpose.

Further difficulties were encountered, however.
The district subcommittee on the Seine Basin report-
ed on staff and technical weaknesses of the dams-and-
reservoirs service in the Seine prefecture. The
Seine Technical Directorate of Water and Sanitation
then agreed to reinforce the dams service, and the
aid of Eléctricité de France was solicited for
engineering studies.

Although 45 per cent financial participation by
the national government in the capital cost was
agreed upon in the district plan (and given for earlier construction projects), the budget directorate
of the Ministry of Finance declared (in opposition
to advice of a committee of the Economic and Social
Development Fund) that national participation could
not exceed 30 per cent. (This position was based on
the theory that the project was exclusively for
flood protection of inhabitants of the immediate
area.)

At this point, the district subcommittee issued
a statement defending the urgency and general interest of the project and insisting on 45 per cent participation by the national government. The cabinet
of the Ministry of Public Works defended this position in a letter to the Ministry of Finance in April,
1963. Under urging from the delegate general of the
district, the prime minister supported this stand,
and agreement on it was finally achieved in the
Interministerial Committee for the Paris Region.
Construction will begin in 1967, four years later
than the date set by the regional plan and eight
years after the project was proposed by the departmental service.

MASS URBAN TRANSPORTATION

Mass transportation has been regionalized in
the Paris area since 1948. The city subway system
(the Métro) had been built and operated up to that
time by the City of Paris, and major bus services
were private and municipal throughout the region.
After the war, the national government was moved
by the scale of reconstruction problems and by
interservice competition to create the Régie Autonome des Transports Parisiens (RATP) to provide
rapid transit and bus services in the urbanized
area. The RATP took over the subway and bus systems in the City of Paris and bus services in the
immediate suburbs.

The RATP is legally an independent public
authority of a commercial nature with operational
jurisdiction in the Paris Transport District, the
boundaries of which approximate those of the agglomération of Paris.* It is directed by an intergovernmental board of twenty members: the president, appointed by national decree; five representatives of local authorities (two from the Paris
council, two suburban representatives from the
Seine council, and one named by a mixed committee
from the départements of Seine-et-Oise, Seine-et-
Marne, and Oise); four national government representatives appointed respectively by the ministers of
interior, finance, construction, and public works
and transport; five representatives of the RATP
employees and officers nominated by staff organizations and appointed by the minister of public works
and transport; and five persons appointed by that
minister for reasons of special competence. The
board appoints a secretary-general, but the executive director of the RATP is named by ministerial

---

*The Paris Transport District includes all of
Seine, a large part of Seine-et-Oise, and several
communes in Seine-et-Marne--a total of some 620
square miles, in which some 7.5 million persons re-
side.

decree on proposal of the board president. Top
personnel are in the State civil service and the
remainder (to a total of 35,000) are legally
employees of the nationalized economic sector,
regulated by special statutory provisions similar
to those for the civil service.

National-government control is thus inherent
in the structure of the RATP, which is further
subject to operational controls--the government
approves its budgets and determines fare structures
and salaries. Administrative supervision is exercised by the Ministry of Public Works and Transport,
which has approval powers over specified decisions
of the board. The transport director in the ministry attends board meetings. The Ministry of Finance
posts a comptroller within the RATP, and its accounts
are scrutinized by the Court of Accounts as well as
by the Commission for Verification of Public Enterprise Accounts. General policy supervision is exercised by the Syndicate of Paris Transport within the
Ministry of Public Works and Transport, while systems operations have been supervised by the Directorate of Bridges and Roads in the Seine prefecture.

The RATP provides all the rapid-transit service--
the subway in the city and one suburban line to
Sceaux--and major bus services within the city and
suburbs. Its systems carry 75 per cent of persons
traveling to work in the transport district by all
modes of transportation.

Commuter railroad services are provided in the
Paris Region by the nationalized corporation that
has a monopoly on all railroad services in France:
the SNCF (Société Nationale de Chemin de Fer). Its
system, which fans out from Paris throughout the
nation, provides intraurban commuter service on
some twenty lines, the management of which is not
separated from general management within the corporation.* According to the general rules governing

---

*Management of the suburban railroad services

nationalized industry in France, the SNCF is subject
to strong controls as to both operations and finance, and the government appoints the majority of
its board.

In addition, major communes in the region outside the transport district have municipal bus services that they operate directly (en régie) or by
concession to private corporations. These and some
private suburban companies belonging to the highway-
transit professional association altogether operate
about fifty local lines.

Finally, the Syndicate of Paris Transport is a
unique intergovernmental syndicate created in 1959
both to concentrate government supervision and regulation of transport in the region and to facilitate
technical and economic coordination of services provided by the RATP, the SNCF, and the small companies--
particularly to harmonize RATP and SNCF fares and
service schedules. Its jurisdiction is the transport district.

Previous to the creation of the syndicate,
the Regional Office of Paris Transport in the Ministry, which also included an intergovernmental committee, had been responsible for coordinating the activities of the transport agencies in Paris, but had
little impact on the independent hierarchies controlling the RATP, the various SNCF divisions, and
small companies. This situation demonstrates the
common conditions under which intense national controls do not produce cohesive administration in
Paris.

---

in the region is fragmented among four divisions of
the SNCF, by virtue of the fact that each major line
that radiates from the City of Paris belongs to a
separate regional division of the SNCF. Each of
these divisions enjoys considerable operating autonomy.

The syndicate is managed by a board of directors of seven members: three appointed by the ministers of public works and transport, finance, and interior; three representing local authorities, one from the City of Paris, one from suburban Seine, and one from the other départements; and the president of the board, who is an administrative officer appointed by government decree. The director of land transportation from the Ministry of Public Works and Transport attends board meetings. The top management officials consist of a secretary-general and technical staff.

Although it is structured like an intergovernmental special authority, this organization functions more like a ministerial agency for policy coordination and supervision. It assumed the responsibilities formerly lodged in the Regional Office of Paris Transport, as well as some supervisory powers formerly resting with the prefects. In fact, it is housed in the ministry offices and its operating expenses are covered by the ministry budget. The minister or government may overrule decisions by the syndicate board or staff.

The syndicate has greater powers than its predecessor to establish the policy framework within which transport in the Paris area operates. It approves creation of new lines and systems, and their manner of operation, in which capacity it is charged with assuring conformance with long-term development plans of the District of the Paris Region and of General Planning Commission.   In theory, the syndicate engages in transport planning of a greater degree of specificity than these bodies but in harmony with their proposals. The syndicate regulates transport operations by fixing the fares of the RATP lines and the SNCF suburban routes and by defining the general conditions of RATP service operations, which are currently set forth in an agreement between it and the RATP . The staff examines the operating accounts and balance sheets, budget

proposals and service plans, work programs and acquisition plans of the RATP. The transit agency still
reports directly to the minister of public works and
transport, however, but he utilizes the syndicate in
a staff role to analyze its annual reports and exercise major approval powers.

In general, the Syndicate of Paris Transport
represents an attempt to concentrate government controls of transit and commuter rail services formerly
scattered throughout various national offices to
heighten coordination and plan-implementation possibilities. The representatives on its board reflect
the nature of transport administration in the region,
which might be summed up as intergovernmental adminis-
tration--organized regionally and dominated by the
national government. To date, however, it has been
engaged in lengthy negotiations with the transportation agencies, whose tendencies to independent action
and recourse to various ministerial allies persist.
Recently, the regional prefect has been given special
powers to coordinate transit projects for purposes of
plan implementation.

Operating Finance

Both the RATP and the SNCF services are subsidized. The RATP must submit an annual balanced
budget with suggested fare structure. If the syndicate or the minister fails to approve the proposed
fare structure, generally by failing to approve a
fare raise, public subsidy is allotted to the RATP
to cover resultant deficit in the operating budget.

In 1963, 61 per cent of the RATP operating receipts of 1,155 million francs were derived from
fares; 33 per cent from government subsidy to com-
pensate for extension of fare-reduction privileges*
and refusal to raise fares on request; and 6 per

---

*Such privileges are extended by the State in
France to virtually hundreds of categories, from
disabled veterans to university professors.

cent from various other sources. The subsidy is
allocated by the syndicate according to formula:
70 per cent is charged to the national government,
10 per cent to the City of Paris, and 20 per cent
to départements served. The operating subsidies
to the SNCF are allocated in a similar manner, the
local participation in both subsidies constituting
obligatory expenditure.

Fiscal management of both the SNCF and the RATP
is complicated by the fact that both fares and personnel wages are fixed by national authorities on
the basis of criteria not related to the transit
operations. Financial policy, in which the Ministry of Finance, of course, plays a large role, has
kept both of these factors low. Transport fares
are elements in the cost-of-living indexes published
by the government that serve as a basis for various welfare allowances and as a measure of inflation, which the government is attempting to control. Salaries in the public sector, which are
uniform, have been held down in spite of a crisis
climate that inhibits both personnel productivity
and recruitment. The recurring RATP strikes of
1962, 1963, 1964, and 1965 are symptomatic of this
problem.

Responsibility for Highways

There are national, departmental, and local
highways and roads in the region. The departmental
bridges and roads services (Ponts et Chausées) construct and maintain national and departmental highways; the Technical Directorate of Paris Streets,
a municipal service situated in the Seine prefecture and operating under the City of Paris budget,
is responsible for street and highway construction
and maintenance in the city. To the latter has
fallen the job of constructing a peripheral highway around the city.

The bridges and roads services in Seine-et-
Marne and Seine-et-Oise operate as field services

of the Ministry of Public Works and Transport (under
the national budget) when they deal with national
highways and as departmental services (under the
département budget) when they deal with <u>département</u>
highways. They also provide, at cost to the communes, construction and maintenance services for
municipalities that do not have their own road services.

The highway function in Seine is regulated by
special statute. There are two branches of the
bridges and roads service: One is field service
of the ministry and the other, a departmental directorate with separate personnel and budget. The prefect is chief executive with respect to both, as
agent of the national government in the first instance and as agent of the departmental council in
the second. He is also, of course, chief executive
with respect to the Paris street directorate, acting then as agent of the city council. The first
branch of the Seine service is responsible also for
parking facilities of regional importance, such as
those at the end of mass-transport lines.

The bridges and roads services are in the process of reorganization. Generally, they have been
structured in three levels within the <u>département</u>:
(1) a directing and staff division (engineering,
personnel, financing, etc.); (2) subordinate
offices with territorial bases; and (3) smaller
territorial subdivisions that perform construction
and maintenance activities. This form, together
with the divided Seine services, has proven too
complex and subdivided for efficient undertaking of
large-scale works. Simplification and consolidation is therefore under way.

New national and <u>département</u> highways are almost entirely financed by the national government
through a special highway-investment fund and the
economic-development fund. The priority given to
development of a national-highway system, which is
glaringly absent in France, accounts for the special earmarking of funds. Highway development in

the Paris urban area, however, is linked to mass-
transport development at three points in the governmental processes.

First, the Ministry of Public Works and Transport both controls the highway services and supervises the transport agencies. As field services,
the bridges and roads units are subordinate, theoretically through the prefect, to a corresponding
division in the ministry. (This situation exempli-
fies the difficulties of the prefect's coordinating
task. The technical corps of civil engineers staffs
the services at both the ministry and the departmental levels; esprit de corps, as well as shared competence in the technical aspects of the service,
limits severely the effective intermediary role
played by the prefect.) In all cases, the ministry
has powers of prior approval over periodic work programs (tutelle à priori) and inspection controls over
work done (tutelle à posteriori). Major works require authorization by ministerial order, while
prefectoral order suffices for others. However,
different directorates of the ministry supervise
highways and transit. Each tends to ally itself
more with its subordinate agencies than with general
ministerial authority.

Second, highways and transit are subject to
consideration in the same planning efforts; they are
major considerations of the district plans. All
highway projects are supposed to conform to the
regional land-use plan; and in the process of interagency review, the regional prefect must be consulted on this point.

Third, the major aspects of both highway and
transit projects for the Paris Region have engaged
the attention of the Interministerial Committee for
the Paris Region, the Council of Ministers, and from
time to time the President of France. Even though
the highway function within the City of Paris is
legally the responsibility of the city, the national
government has made such detailed decisions as those
on parking-facility development for the city and the

routing of the peripheral highway. The face of
Paris is the concern of France; while there have
been periods of neglect, the national prerogatives
remain jealously guarded.

Appraisal of Current
Transport Facilities

Virtually no major investments in mass-transport systems in the region were made for almost
fifty years, while the population grew by over 40
per cent, mainly in the outer portions of the
agglomération (some communes have grown by 300 per
cent), and jobs remained concentrated in the center.
That the system functions as well as it does is
testimony to the farsightedness of its original
designers. The current problem--a strongly felt one
among urban problems--is that of the journey to work,
and the current need is for major capital investment.* The average daily travel time of 2,300,000
commuters is one and a half hours round trip. Almost half a million change modes (to say nothing of
vehicles) once or twice in their journeys to work
(see Table 7). Peak-hour transportation is locally
considered one of the most severe urban problems on
the basis of both passenger complaints and rapid
growth of automobile traffic, which the City of
Paris is unsuited to absorb. One need only watch
seven lanes of traffic from three directions funnel-
ing through a two-lane arch of the Louvre to under-

---

*Total RATP traffic has been fairly steady in
the postwar period, with small declines since 1959--
partially attributable to work stoppages--somewhat
offsetting increases before 1959. But peak-hour
rapid-transit traffic rose 12 per cent from 1953 to
1963. SNCF Paris traffic has been growing rapidly
(by 16 per cent between 1953 and 1959). Motor-vehicle
traffic is outpacing them all, however, with registrations in Seine alone having increased by 40 per
cent between 1953 and 1959. In the 1960's the number
of automobiles in the region has been increasing by
about 5 per cent per year.

stand the significance of former councilman Griot-
eray's observation: "Paris has existed for 2,000
years and the automobile for 50!" The City of
Paris, with its limited financial resources, must,
nevertheless, cope with the inadequacies of internal highways and parking facilities outdated by the
rapidly growing commutation by automobile.

Transit services within the city, however, are
good; the density, speed, and frequency of the Métro,
the low fares, and the experimentation (e.g., with
silent rubber-tired cars and a moving pedestrian
belt at an underground connection) are notably
positive features. But the subway traffic is at
the saturation point during rush hours on major
lines, and city bus services are becoming slow and
undependable because of traffic congestion. Passengers and local politicians complain that the RATP
is isolated from public pressure and operated not
as a public service but as an economic enterprise.

The deficiencies of RATP suburban bus ser-
vices--slowness, infrequency, and discomforts of
old equipment are the major complaints--leave the
slices of suburban land between the railroad lines
poorly served. The most severe deficiency is in
highways, however, and this limits the present capacity to improve the suburban bus systems. There
are only two major short highways into the city
completed to date--to the south and west.* Moreover, railroad service has not been fast and fre-
quent enough to satisfy commuter wants.

The government units and agencies most interested in improving passenger transport facilities in
the Paris Region did not have the financial resources to undertake improvements in the postwar period;
the restrictions on borrowing and general fiscal
resources of the city, the départements, the communes and the RATP,together with the investment

---

*One to the north and the inner-belt highway
are under construction.

priorities and decentralization policies of the
national government (which did control the resources),
go far to explain the existing situation.

Planning and Investment Programs
for Mass Transportation

As was true for water supply, the effective
project plan for transportation at present consists
in the relevant sections of the district four-year
plan, which utilized prior planning work from the
1960 land-use plan, the national four-year plan,
and project plans of various operating agencies.
The district twelve-year and new land-use plans set
transportation-investment programs into the context
of general development goals for the region.* At
this point, planned programs are designed to meet
obvious current needs and projection of present
traffic-growth trends, in light of the general
policy expressed in the four-year plan. This policy
holds that, although large-scale highway development
is essential to meet rapidly growing automobile
traffic, the regional system cannot function efficiently without complementary growth of mass-transit
facilities and linkages between these and highway
routes.

A single work group on transport, traffic, and
parking prepared both the highway and the transit
sections of the four-year plan. Headed by an official of the bridges and roads corps (the president
of the High Council on Transport, a national interagency committee), its membership was derived primarily from this corps, the operating transit agencies, and the urbanism service of Seine, with one
representative of city administration--the general
director of technical services for the City of Paris.

_____

*The regional land-use plan calls for ten
radial highways and a second belt system in the
suburbs.

The whole plan as approved emphasizes the
priority of transportation projects, a priority
that was strongly felt by the district council as
well as by the district staff,* but the plan cites
this as the one area in which financing problems are
extraordinary. The plan proposes bus-service improvements including purchase of new buses and creation
of some new lines, and urges speeded scheduling of
previously proposed highway projects. It adopts
proposals for improvement of SNCF commuter service
from the national four-year plan and adds one new
SNCF program. Total investment by the State and
SNCF would be 597 million francs for electrification
of several suburban lines, construction of a commuter
service station underneath the Gare d'Austerlitz,
various system improvements, and (this added by the
district) construction of an interchange station
linking the SNCF lines to the new rapid-transit
express at Rond Point. Reorganization of the SNCF
budget to separate Paris commuter from national service is recommended on grounds that SNCF action on
commuter-service problems would be more likely to
be forthcoming under such an arrangement.

The plans for expansion of rapid transit include line changes, elongation of trains and stations, purchase of rolling stock, elongation of two
lines, and--the most important project in the four-
year program as a whole--starting work on the regional transit express (RER). Total capital investment proposed for the plan period on these transit
projects is 2,112 million francs. Of this, 26 per
cent is to be financed by the national government,
26 per cent by the district, and 48 per cent by the
RATP through loans and grants. (By comparison, investment in the plan period proposed for highway
programs is 4,908 million francs.)

---

*In fact, the parliamentary debates on creation
of the district in 1961 emphasized the expectation
that it would give priority to the region's mass-
transportation system.

TABLE 7

WORKING PERSONS BY MODE OF TRAVEL FOR JOURNEY TO WORK
IN THE PARIS TRANSPORT DISTRICT

Mode Used	Division by Modes Used	Per Cent
ONE MODE ONLY USED		
Railroad	112,300	4.1
Métro (rapid transit)	524,200	19.0
RATP bus	286,000	10.4
Private bus and taxi	7,100	.3
By employer	23,900	.9
Automobile	271,200	9.9
Bicycle, Scooter, etc. ("2 wheels")	285,300	10.4
TWO MODES USED a		
Railroad and Métro	140,300	5.1
Railroad and RATP bus	40,600	1.5
Métro and RATP bus	211,500	7.8
Railroad and one other	14,400	.5
Métro and one other	14,600	.5
RATP bus and one other	5,300	.2
Two others	2,000	-

Mode Used (continued)

	Division by Modes Used	Per Cent
THREE MODES USED[a]		
Railroad, Métro, and RATP bus	16,300	.6
Railroad, Métro, and one other	7,900	.3
Railroad, RATP bus, and one other	1,700	.1
Métro, RATP bus, and one other	1,800	.1
Railroad and two others	200	-
Métro and two others	Negligible	-
RATP bus and two others	100	-
Three others	Negligible	-
TOTAL TRANSPORT USERS	1,966,700	71.7
TOTAL USERS OF NO MEANS OF TRANSPORT	774,400	28.3
TOTAL WORKING PERSONS	2,741,100	100.0

Source: Direction de la RATP.

[a]"Other" refers to Private bus and taxi, By employer, Auto, and "2 wheels."
Divisions between these four are estimates.

151

District proposals for long-term allocation of
capital financing for mass transportation focus almost entirely on creation of the regional express
system and place the total capital burden ultimately
on the national government and district budgets.

The Regional Transit Express

The regional transit express (RER) is to be a
system of rail express service that will link some
existing railroad lines, existing subway lines,
several new rapid-transit lines, and new railroad
connecting branches. As planned, it will cover the
transport district and tie into the railroad lines
to outer portions of the region (see Map 4). Construction of new east-west and north-south rapid-
transit express lines in the city (via deep underground tunnel), transfer stations and SNCF-RATP
linkages, and electrification of SNCF commuter lines
are required to create the first phase of the system. The east-west transit line, scheduled for completion by 1975, is the first component. It will run
from St. Germain-en-Laye, on the western border of
the agglomération, to Boissy-St. Leger, to the east.
Construction is under way on part of it: a new
rapid-transit underground line from La Folie to
Vincennes running under the heart of the city and
having six stops within it (one new station and five
existing Métro stations). The east-west line is
expected to carry 600,000 to 800,000 passengers per
day and is designed to discourage auto travel into
the dense urban center. Each suburban station will
have free-of-charge parking facilities. Fares will
vary with distance.

The project gradually took shape through several planning efforts, each of which entailed consultation with the operating agencies and official government approval. The 1960 general land-use plan
(PADOG), prepared by the Ministry of Construction
with the aid, in this case, of the Ministry of Public Works and Transport, had mapped a proposed regional express system. It pointed out:

In a few years commuter travel into the center of the Paris Region will be considerably hindered by paralysis of different modes available today. The creation of a regional rail express system is an overwhelming necessity. Only a mode of transport on a regional scale with high speed superimposed on existing systems can bring real improvement in the living conditions of the population.

Rapid-transit expresses linked to SNCF rail lines and RATP express bus services were proposed, as were details of the east-west line that are close to those of the present project.

The fourth national plan prepared by the General Planning Commission (the first national plan to treat intraurban transportation) accepted the principles of PADOG by setting forth as a target for the plan period (1962-65) commencement of work on the east-west trunk of the regional system.

After publication of the plan, the Paris Region Institute for Planning and Urbanism conducted more detailed research on commuter traffic in the Paris Region. Its conclusions were the basis for the district four-year-plan proposals, which stressed that a speeded work schedule for the east-west trunk was required (stating, for example, that saturation of the Métro line number 1--to the west--would come in 1969, not 1973, as the national plan envisioned). New cost estimates were presented. Table 8 shows the investment proposals relating to the east-west trunk from both the national and district plans for 1962-65, illustrating the degree to which the national plan serves as a flexible framework for regional planning.

The work by the SNCF related to the express system is financed by borrowing and national subsidies, requiring specific approval by the railroad's board and national finance authorities. Most of the work on the east-west trunk, however,

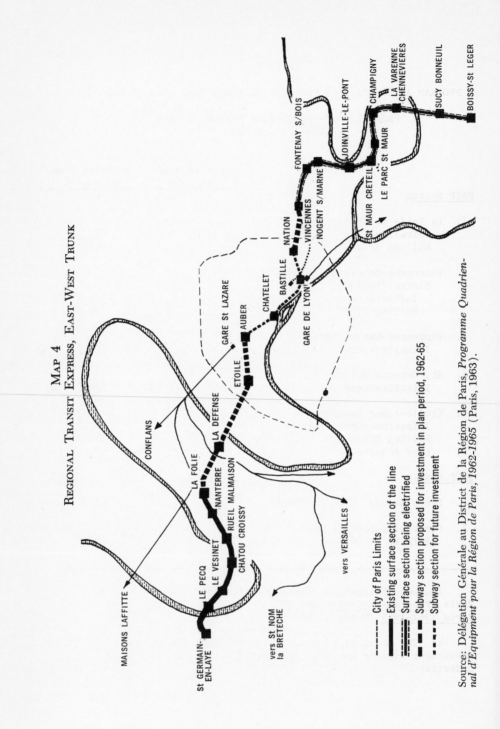

MAP 4
REGIONAL TRANSIT EXPRESS, EAST-WEST TRUNK

City of Paris Limits
Existing surface section of the line
Surface section being electrified
Subway section proposed for investment in plan period, 1962-65
Subway section for future investment

Source: Délégation Générale au District de la Région de Paris, *Programme Quadriennal d'Equipment pour la Région de Paris, 1962-1965* (Paris, 1963).

TABLE 8

PROGRAM AUTHORIZATIONS FOR EAST-WEST REGIONAL TRANSIT LINE

PROPOSED FOR FOURTH PLAN PERIOD, 1962-65
(million francs)

	National Plan Proposals	District Plan Proposals
RATP System		
1. La Folie to Etoile Section:		
Infrastructure	390.0	226.0
Rolling Stock	47.0	52.0
2. Vincennes-Boissy-St. Leger, Electrification:		
Infrastructure	100.0	171.0
Rolling Stock	123.0	134.5
3. Vincennes-Nation Section:		
Infrastructure	-	228.0
4. St.-Germain Branch:		
Infrastructure	-	120.0
5. Etoile-Auber Section:		
Infrastructure	-	155.0
Rolling Stock	-	106.0
Land Acquisition	-	50.0
SNCF System		
1. Paris-Nantes, Electrification	- [a]	222.0
2. Connecting Station with RATP at la Défense	-	13.8

Source: Délégation Générale au District de la Région de Paris, Programme quadriennal d'équipement pour la région de Paris 1962-1965 (Paris, 1963).

a. The national plan did not break down the proposed investment figures for this work but did provide that it be undertaken.

is that to be done by the RATP. The district plan
proposes that the district itself finance 533 million francs of it; the national government, 533 million francs; and the RATP, the remainder. This allocation had been worked out through negotiation and
agreement in 1961-62.* Authorizations required thus
include those from national and district budgets
(with national approval) and RATP budget authorizations (with approvals by the Syndicate of Paris
Transport and the government). The authority of
the Ministry of Finance is significant in all three
cases.

The district plan pointed out that 533 million
francs (the largest district allocation by far for
any work) would take all district resources for the
plan period if it did not borrow (and proposed later
commitments would surpass them). It further pointed
out that RATP financing for this and other plan projects required considerable augmentation of its resources, suggesting fare raises, long-term low-
interest national-government loans (such as
those given to housing authorities) and national
grants.

Early decisions had been taken by the RATP before publication of the plan authorizing work on
the first two sections of the trunk, and related
expenditures were included in the 1961 and 1962 RATP
investment budgets. In 1963, at the request of the
district, changes in routing and corresponding investment authorizations for the project were made by
the RATP. Further authorizations were included in
the 1963 RATP investment budget. In 1962 and thereafter, the RATP received national and district grants

---

*The 50-50 sharing of direct government grants
for the project between the district and the national government was first suggested by the Economic
and Social Development Fund.

for the work.

These authorizations, however, were too low to
meet the rhythm of work contemplated in the district
plan. The district is urging the national government to step up its financing of the project, and
in the district 1964 budget proposal, it requested
approval of district borrowing to meet its own
obligations on schedule.

The events that followed are described in
Chapter 3 (Pages 78-79). In order to assure long-
term financing of the regional express on schedule,
the district council, on advice of the delegate
general, had voted separate receipts and expenditure from long-term borrowing by the district of
457 million francs for grant to the RATP. The
Ministry of Finance, however, prohibited earmarked
borrowing for the grants to the RATP and ruled that
district borrowing would have to be authorized
annually and scaled to meet deficits in payment
schedules of the over-all district grant program
in each given year. The ruling was based on fiscal
policy, particularly stabilization policy, and government support of it was couched in terms of national interests taking priority over regional interests.
District budget authorizations for the regional
express were maintained (524 million francs were
budgeted for the project by the district in the
plan period, compared with 533 proposed in the plan),
but other district programs were reduced and the
payment schedule slowed down. The total borrowing
by the district authorized for the plan period was
70 million francs.

On balance, then, the efforts by the district
to assure in advance investment in the east-west
trunk at a specified rhythm were not successful.
The work remains contingent upon annual fiscal decisions. Nevertheless, the district efforts have
definitely increased the credit available for the
work, and its prodding of the government to make
favorable fiscal decisions at each decision-making
stage continues.

The implementary construction work on the trunk
line, mainly by the RATP, has been slowed somewhat
by technical difficulties of underground terrain at
the tunnel site under the center part of the city.
Study of rolling-stock types for the service and
construction of stations is under way by the RATP;
the SNCF is working on the line and stations between Vincennes and Boissy-St. Leger (the costs of
this work are charged to the RATP budget).

PUBLIC-HOUSING PROGRAMS

Government housing programs are uniform throughout France; all involve national subsidies to public, private, and cooperative agencies organized and
building according to national law and regulations.
One major category is HLM housing* built by public
authorities established by communes and départements.
HLM housing is also built by cooperative societies.
Together these two types of HLM construction accounted for 30 per cent of all new dwellings built in the
Paris Region in 1963.

Other arrangements for subsidized housing
accounted for about 60 per cent of all new dwellings
built in the region that year. These are constructed mainly by private agencies--building societies,
cooperatives, limited-profit corporations--but also
by mixed corporations in which local-government
capital participates. The City of Paris is the
majority shareholder in RIVP, for example, which is
a mixed corporation engaged in urban renewal and

---

*HLM denotes Habitations à Loyer Modéré, or
moderate rental housing as defined in housing legislation. It also refers to the local-government
agencies and the cooperatives organized to construct
HLM housing. These HLM organizations also build two
other kinds of housing under State subsidies: "nor
mal rental housing" and low rent "social housing,"
both precisely defined by law.

large-scale housing construction in the city. The
other shareholder is the Bank of Paris. Interest
subsidies, extended through loans from the Caisse
des Dépots and Crédit Foncier,* and ten-to-twenty-
year annual-building premium grants are extended
by the national government.

The remaining proportion of new dwellings
built--10 per cent--were unaided, primarily single-
family houses. The size of this sector reflects
the extent to which the economics of housing have
rendered rental housing unattractive to private
capital in the region.

HLM Public Authorities
in the Paris Region

The HLM public authorities are specially established by communes, départements, and intercommunal
special districts and operate under autonomous
management. Creation of an HLM authority requires,
after local council vote, approval by joint decree
of the ministers of interior and construction.
These ministers must consult the national and departmental advisory commissions on public housing.
There are 49 HLM public authorities in Seine (1
departmental authority, 1 City of Paris authority, and 47 other municipal authorities), 7 within
Seine-et-Marne, and 8 within Seine-et-Oise. Their
operating expenditures and initial capitalization
are voted by the councils that have established
them from the local budgets. The Directorate of
Construction within the Ministry of Construction is
the major supervising authority over the HLM agencies.

---

*Caisse des Dépots is a publicly owned bank
that manages the funds of savings banks (convert-
ing short-term savings into long-term credit);
Crédit Foncier is a mixed-corporation mortgage
bank.

The HLM agencies build rental housing capital-
ized by low-interest loans: 1 per cent loans with
forty-five-year amortization periods granted by the
Ministry of Construction, to cover up to 85 per cent
of construction costs, plus 5.25 per cent supplementary loans from national banks for the remaining 15
per cent of construction costs. The ministry con-
sults a special interministerial commission on loan
approvals. In addition, the agencies build "normal rental housing" with the aid of thirty-year, 5
per cent national-government loans, and low-income
housing with no-interest fifty-three-year loans.
Land and supplementary loans and grants are extended to them by the local councils.

The Seine HLM authority has constructed about
50,000 dwelling units, and its present project plans
include 8,000 more. The Paris authority has completed
some 55,000 units, some of which are located outside
the city, where more reasonably priced land could be
obtained.

HLM authorities enter into contracts with private or mixed corporations for construction and
manage their housing projects directly. By delegation from the ministry, their functions can be expanded to include urban renewal, land development,
and other construction duties. (The City of Paris
authority is engaged in urban-renewal operations.)

That land granted to the authorities, generally
by the local councils, can be acquired through the
departmental land services (field services of the
Ministry of Construction) or through the regional
land agency (see Chapter 4) lifts the burden of
expropriation proceedings from the HLM authorities
and the local administrations.

Each HLM authority is managed by a board of
twelve directors, of whom four are appointed by the
creating municipal or departmental council; two (one
each) by the national housing funds; and six by the
appropriate prefect. The chairman is usually mayor
or assistant mayor in the case of a municipal

authority, or a departmental councilor or local
member of Parliament in the case of a departmental
authority. The boards utilize five-man subcommit-
tees to make decisions on allocation of HLM dwelling
units, with the exception that this function has been
handled in the City of Paris and Département of Seine
by central authorities since 1963. The removal of
this power from local authorities was met by intense
hostility on the part of local officials in Paris.

Tight operating control of the HLM authorities
is maintained by the central government through both
administrative and financial supervision. Board
decisions respecting budgets, acquisitions of buildings, and transfer of stocks and shares require approval of the prefect, who also appoints a supervising
commissioner for each authority. Financial controls
include audit, inspection, and posting by the Ministry of Finance of a comptroller to handle the actual
transfers by the housing authorities. Housing-
project plans must be approved both by the local
councils and by the Ministry of Construction. In
addition, the local council exercises general supervision of operations of a housing authority that it
has established.

Departmental advisory committees on HLM housing, which are appointed by the prefects to coordinate and to stimulate the activities of the various
authorities in the départements, report annually on
these activities to the national advisory committee
attached to the Ministry of Construction. At both
levels, the advisory committees are composed of
representatives of various planning and operating
agencies and of ministerial services.

In addition, housing bureaus of Seine and of
communes and intercommunal special districts in the
other départements undertake housing exchange and
brokerage activities and participate in the National Housing Exchange for HLM and all other types of
housing in the region.

HLM Housing

Building types, standards, costs, and tenant
eligibility of the three major categories of housing built by the local HLM authorities are regulated by national code. The low-income projects eligi-
ble for no-interest loans are restricted to rehous-
ing people living in slums or in buildings being
demolished. Ceilings on incomes for tenants of
moderate rental housing vary with the region and
size of family. For a household of four with one
wage earner, maximum monthly income for moderate
rental housing is 1,929 francs in the Paris Region.
Since the average monthly income of a blue-collar
worker in France is 717.8 francs, this housing is
in effect aimed at middle-income groups. However,
assistance to persons with very low incomes and to
families with children is provided directly to the
families by national housing allowances. Local
authorities are obligated to contribute to the rent
allowance paid to the lowest-income groups and the
elderly.

Nevertheless, HLM rents are about 75 per cent
below those in privately constructed dwellings,
ranging from 80 to 250 francs per month, depending
on size of unit. Rents are based on costs of construction, which are regulated according to housing
type, size, and region of the country. Normal rental housing does not have income eligibility ceilings, but its cost cannot exceed by more than 20
per cent that of comparable moderate rental housing.

The maximum floor space in moderate rental dwellings ranges from 20 square meters for one bedroom
plus toilet and cooking facilities to 125 square
meters for seven rooms plus kitchen and bathroom.
Overcrowding is not permitted in tenancy of these
buildings.

Housing Problems in the Paris Region

Specialized studies, the press, and inhabitants
of the region refer to the housing "crisis" in the

Paris urban area. The problem is both a qualitative
and quantitative one, to which solutions are complicated by land law, land price structure, and rent
controls. Moreover, it differs in kind among sections of the urban area. In the city, the age and
amenities of existing buildings and the scarcity and
price of land are at the root of high unfulfilled
demand for new dwellings at reasonable rentals. The
ring immediately around Paris consists of the poorest
and most substandard housing in the region. After World
War I, suburbs just beyond this ring were subject to
intensive individual home building that was stimulated by government mortgage aids. Here, the major
problems are those of remaining deficiencies in public facilities. At the fringes of the agglomération,
total construction has simply not kept up with population growth.*

Substandard housing in the region is defined by
criteria of age, crowding, and amenities. Of the
total housing stock in 1962, 55 per cent of the dwelling units were constructed before 1915, 29 per cent
between 1915 and 1948, and 16 per cent from 1948 to
1961. The amenities in the existing dwellings are
closely related to age. An advance report for the
district twelve-year plan estimated that 3 per cent
of the dwelling units in the region do not have electricity, 15 per cent do not have water in the dwelling or on the same floor as the dwelling, 45 per cent
have no internal toilet, and 60 per cent do not have
central heating.

The 1962 housing census of France classified 10
per cent of the dwellings in the Paris Region as
critically overcrowded, 15 per cent as having temporarily permissible overcrowding, 50 per cent

---

*While urban housing is a severely felt prob-
lem, it should be kept in mind that in qualitative
terms (particularly age and amenities) rural housing is far less well off (see, for example, Table 9).

TABLE 9

PER CENT OF PRINCIPAL DWELLING UNITS
OLD AND OVERCROWDED

	Built Before 1871		Overcrowded	
	Urban	Rural	Urban	Rural
Seine	18	-	12	-
Seine-et-Oise	13	50	12	14
Seine-et-Marne	23	55	11	12
Paris Region	17	53	12	13
France	22	52	12	15

Source: Ministry of Construction.

TABLE 10

HOUSING CONDITIONS OF MIGRANTS AND ADULTS 50 TO 60 YEARS OLD
IN THE PARIS AGGLOMERATION

| | Migrants[a] | | Persons Aged 50 to 60[b] | |
	Recent (Per cent)	Arrived Prior to 1945 (Per cent)	Born in Provinces (Per cent)	Born in Paris Area (Per cent)
Average Number of Persons per Room	1.32	1.14	1.01	.99
Degree of Overcrowding				
1. Critical overcrowding	19.0	12.0	6.0	7.0
2. Temporarily acceptable	21.0	17.0	16.0	13.0
3. Normal	48.0	51.0	55.0	58.0
4. Underinhabited	12.0	20.0	23.0	22.0

Source: Guy Pourcher, Le Peuplement de Paris (Presses Universitaires de France, 1964).

a. Of those in housing classified as critically overcrowded, 21 per cent of recent migrants and 36 per cent of pre-1945 migrants considered their housing acceptable or satisfactory.

b. Of those in housing classified as critically overcrowded, 45 per cent considered their housing acceptable or satisfactory.

as normal, and 25 per cent as underinhabited.* Other
urban areas in France of over 100,000 persons had
about 7 and 9 per cent in the first two categories
and 39 per cent in the last; hence, housing is in
shorter supply in the Paris area than in other urban
areas. The average number of persons per room in
the agglomération is 1.07, while the average number
of persons per room in those dwellings designated
as overcrowded is 2.45. Hence, the pattern of distribution of housing is a factor of the shortage.**

Tables 9 and 10 summarize housing conditions,
showing that recent migrants bear the brunt of the
crisis (as do newly formed families, for which there
are not, however, separate statistics) and that the
situation in the region is nonetheless better than
that of the nation as a whole, which is dominated by
rural conditions.

---

*Critical overcrowding is defined as 3 persons
in a 1-room dwelling, 4 in a 2-room dwelling, 6 in
a 3-room dwelling, 8 in a 4-room dwelling, etc.
Temporarily permissible overcrowding is 2, 3, 4, 5,
and 7 persons respectively in the ascending number
of rooms.

**Rent control has been cited as causing under-
utilization of older dwellings as well as discourag-
ing new building. The district report for the
twelve-year plan, however, found underutilization
to be spread equally between rent-controlled and
postwar buildings. There is no question, however,
that the controls distort the housing market: A
five-room dwelling in a controlled building (the
vast majority of buildings in the central city)
costs about 300 francs per month, while comparable
new units cost between 2,000 and 3,000 francs per
month. "Foot in the door" payments to landlords in
controlled buildings are common and high. Semi-
annual raises in rents in controlled buildings are
so small as to make little reduction in the price
gap.

On the basis of these conditions, the housing census estimated construction needs in the Paris Region at 104,850 dwelling units per year for 1962-72 (of which 98,690 were to be in Seine and Seine-et-Oise) in order to bring housing conditions in the region up to those of other urban areas and to keep pace with population growth. The district four-year plan sets a target of 95,000 dwelling units per year. Neither of these estimates takes into consideration rising standards or consumer demands at various cost levels.*

In any case, the actual total construction rate has been well below this but has recently risen. Housing starts expressed in dwelling units in the region were as follows (with HLM housing accounting for about 30 per cent):

1955	57,000
1958	75,000
1960	70,000
1961	65,000
1962	68,000
1963	78,000
1964	89,000
1965	94,000

The rate of program authorizations, or capital commitment, for dwelling units in the region was well ahead of housing starts in 1963, indicating that the increase is likely to continue. Table 11

---

*The housing services in the region had an average of about 300,000 requests for housing on their records in 1963.

shows the breakdown of authorizations by type of housing.

TABLE 11

1963 AUTHORIZATIONS FOR DWELLING UNITS
BY TYPE

	Region	City
HLM Rental Housing	32,309	649
HLM Cooperative Society Housing for Ownership	3,375	-
Social Housing	32,023	2,198
Other Publicly Aided Housing	37,628	5,834
Unaided Housing	10,466	2,101
Total	115,801	10,782

Investment and Planning

The pace of public-housing construction in the region has been lower than necessary, then, to catch up with other urban regions and to keep up with population growth. It is almost entirely dependent on the availability of national-government credits and local ability to supply land at acceptable costs. The second factor, supply and cost of land, has not changed radically in the past five years. The rise in the construction rate beginning in 1962 is mainly attributable to government policy to loosen credits in the region. Total receipts of HLM public authorities in the region were 479 million francs in 1960, and 592 million francs in 1962. In the 1964 budget of the City of Paris, total expenditure for housing was 66 million francs, of which 90 per cent was financed by State funds.

The district plan points out, however, that allocations of investment in housing made in the national plan are not sufficient to meet the construction targets in the region. This, of course, is a problem of policy that occurred in relation to water supply and transport as well. The district four-year plan does not include a capital program for housing; since the financing responsibility is so dominantly that of the national government, the district role in working out agreements in this area is small.*

The pace of public-housing programs is contingent upon not only the rate of investment but also, conversely, cost factors, the most important of which is the cost of land in the Paris Region. Uncontrolled land prices, discussed in Chapter 4, contribute to the slow expansion of HLM authority programs, particularly in the densely used area at the center of the region . During the 1964 debate on the construction budget, parliamentary critics within President de Gaulle's party blamed the slow rate of housing construction on failure to control land inflation, the insignificant use to date of the legislation permitting notification of priority zones for urbanization (ZUP's), the extremely complicated structure of national housing aid, and the cumbersome process of construction control--such as the permit-approval procedures. In addition, the profits of the founders of building associations receiving financial aid for housing have come under fire and in 1964 were limited by law to 6 per cent. Finally, insufficient industrialization of the building industry (although France is noted for its use of heavy concrete prefabrication),** high costs of private loans and mortgages on the financial market, and anarchy in the

---

*There has been no regional plan for housing construction. The project plans of each HLM authority are legally required to conform to approved land-use plans. Coordination of programs, however, rests with the advisory commissions and the ministry.
**Less than 0.5 per cent of building firms in France employ more than 300 workers.

rental market have been cited.

The structure of organizations engaged in public housing is not a major source of roadblocks to
housing improvement in the Paris area. HLM authorities are generally recognized as efficient but stag-
nant. The low rate of expansion of their activities
over the postwar period is generally attributable to
factors external to their organization--factors such
as centralization of detailed control, land inflation, local financial weakness, and the lack of
agreed-upon regional housing plans and investment
targets. One additional category of problems will
probably be mitigated by the new physical planning
procedures: those arising from previous failure to
design major public-facilities programs (water, transport, schools, etc.) and housing projects in complementary spatial relationships in the region as a
whole.

PRIMARY AND SECONDARY EDUCATION

Education in France is a national service for
which policy-making authority has been concentrated
at the center--now in the Ministry of Education--
since the reign of Napoleon. The education system
traditionally was subdivided into primary, technical, and secondary schools, which were not successive
steps in a single educational process but relatively
autonomous systems serving overlapping age groups.
The secondary system prepared the student for university studies; the primary system provided students
not destined for higher education with terminal education in collèges, in which age groups overlapped
with those of secondary schools (lycées). The technical system trained skilled industrial and office employees. These systems, reflecting a social strati-
fication that is weakening, entail considerable adminis
trative duplication.

Recent reforms have been aimed at integrating
the school systems, increasing transferability from
one system to another, and strengthening mass general education. The Ministry of Education underwent
several successive reorganizations between 1959 and
1964, during which the traditional autonomy of the
major divisions corresponding to the three school
systems proved tenacious. The ministry is presently
subdivided into functional directorates--such as
school administration and teaching--which deal with
all three systems. A post of secretary-general (not
usually found in national ministries in France) was
created (and a man to fill it brought from another
ministry) to direct and to coordinate the work of
the various divisions. The budget, finance, and
school planning services are attached directly to
his office.

The education system as presently constituted
is divided into cycles. The first cycle, or primary
education, includes nursery schools and elementary
schools of five grades for children six to eleven
years old. Following is the two-year observation
cycle, which is offered by both primary and secondary schools but for which the curriculum is being
made uniform. At the end of this stage, at about
fourteen, the student is directed to suitable higher
education on the basis of judgment by a council of
professors and school directors. (Formerly, at age
eleven, students would go into terminal general education in the primary schools or into lycée courses
preparatory for higher secondary school.)

The second cycle includes four major alternatives: (1) long general education course at a
lycée, lasting until the student takes the bacca-
lauréat examinations for university entrance at about
eighteen; (2) long technical course at a technical
lycée, leading to engineering or scientific higher
education; (3) short, general education course,
which lasts to age sixteen and is terminal, taken
at collèges d'enseignment général, which are attached to primary schools; and (4) short technical
course giving skill training at technical collèges.

Education Field Services
in the Paris Region

The ministerial services are responsible for
planning and building national secondary schools
(the majority of secondary schools in the region);
regulation, direction, and inspection of all
schools; curricula, examinations, and content and
form of all accredited education programs; appointment, payment, and supervision of all teachers.
Policy-making with respect to educational programs
is centralized in the ministry itself. School
administration is effected through regionally and
departmentally organized field services.

The academie is the regional educational district. When the academies were recently modified,
that for Paris was reduced from nine to four départements: Seine, Seine-et-Oise, Seine-et-Marne, and
Oise. The administrative head of the academie is
the rector, who is a direct representative of the
minister and a high civil servant appointed by presidential decree. He is a professor who is concurrently administrative head of the university of
the academie. The departmental branches are directly
responsible to him, with the sole exception in France
of that of Seine, which has its own director, who is
responsible directly to the minister. The rector of
the Paris academie has only supervisory powers over
lycées within Seine. In 1960 and 1961, some deconcentration orders were promulgated, expanding the
role of the rector by transferring functions from
the central ministry.

The rector's own administration includes a
cabinet and bureaus for primary, secondary, and
higher education, each directed by an appointed
teacher. (Technical education has regional and
departmental offices that report directly to the
ministry.)

Secondary education is administered directly
from the academie. "Regional inspectors," subordinate to the rector, act as supervisors and comptrollers

of secondary schools. Primary education is managed
at the département level, under general supervision
of the academie primary-education bureau, which
organizes examinations, controls personnel, and
reviews inspection reports.

The educational services at the departmental
level are headed by officers called inspectors of
the academie in Seine-et-Marne and Seine-et-Oise and
by the director general of education in Seine. These
officials direct primary-school management--nominat-
ing teachers for appointment by the rector, supervising teaching through subordinate primary-school
inspectors (headmasters are administrative heads
only, the primary school inspectors being the direct supervisors of teachers), and working with the
prefect on matters of school construction and facilities. Each primary school inspector has jurisdiction over a limited number of schools and reports to
the departmental inspector of the academie or, in
Seine, to the director general.

The director general of education in Seine combines the roles of rector and departmental inspector.
This consolidation of normally academie and departmental branch functions is justified by the fact that
a large number of top-flight national lycées are concentrated in Seine. He also administers City of
Paris education functions. Hence, consolidation of
national, departmental, and local administration in
Paris and Seine is maintained in education.

Consultative advisory education committees are
utilized at both the academie and the département
levels. The academie committee includes the rector,
ministerial officers, university professors, secondary-school teachers, and two local government representatives. There are several département school
committees on which education officers, teachers,
the prefect, and representatives of local authorities sit. One important committee so composed examines and attempts to coordinate all school building
plans and receives complaints and suggestions from

parents and citizens.

Finally, various corps of general inspectors
from the central ministry oversee education in the
region. One corps has responsibility to examine
school building and organization; another, which is
specialized by subject area, oversees teaching.
These are superimposed over the field-service branches,
and their powers extend to inspection of private
schools.

Département and Communal Roles

The communal councils are responsible for building, equipping, and maintaining primary-school plant
(an obligatory expenditure). Departmental councils
must provide teacher-training schools.

Furthermore, under their general powers, both
communes and départements can establish collèges or
lycées, which coexist with national secondary schools.
Such local schools are managed, directed, and staffed--
as are national schools--by the ministerial services;
the main difference is that they are constructed,
equipped, and maintained by the local authorities.
All teachers are appointed, paid, and supervised by
the national government.*

For Paris and Seine, the local education activities are the responsibility of the respective councils, each of which has an education committee, but
these activities are administered by the directorate
general of education in the prefecture, which, as
has been noted, operates as a national, departmental,
and local service combined. Elsewhere in the region,

----

*The City of Paris and Département of Seine have
developed, on the basis of general powers, special
education programs in such fields as painting, music,
and management, using lay experts as part-time teachers. Recently, however, the national government has
taken action to merge these into the national teaching
corps.

municipal and department school services are under
direction of the mayor or prefect (and their respective secretaries-general) and are separate from
ministerial field services.

For all of these "local schools"--primary
schools and communal or departmental collèges or
lycées--the local authorities must submit building
plans through the prefect to the Ministry of Education where, if they are accepted, they must be in-
scribed in the general education plan before State
aid is available.* National standards of construction and maintenance are enforced through inspection.

National-government aid for primary-school
construction varies with the type of school, but it
averages about one third of costs and is administered
through the prefect. For optional local secondary
schools, the municipality or département is given two
choices: (1) The national government undertakes
construction, and the financial participation of the
local unit is fixed by contract, generally at about
40 to 50 per cent of the cost estimates; or (2) the
local unit undertakes construction, and the national
grant is fixed by contract (usually 50 per cent of
cost estimates), with financial risks falling on the
local authority. The first option is more frequently
taken by local units in order to limit their costs to
a sum known in advance. In any case, most secondary
schools in the Paris Region are national schools.

Teacher Recruitment

Teachers are members of the national civil
service. Primary-school teachers are recruited by

-----

*Theoretically, local authorities could build
schools under their general powers without national
approval and aid; this is an unrealistic alternative,
however, since they would run the risk of the ministry's failing to accredit the school or to assign
teachers to it.

national examination at age sixteen and are subsequently sent through three-year teacher-training courses, during which time they are supported by the government. They can be posted anywhere in the nation. Teachers who are qualified but are not full-time civil servants (e.g., retired teachers, married women serving part-time) are hired under contract to fill shortages.

Secondary-school teachers are recruited by examination after receiving a university degree or completing courses at one of four higher teacher-training colleges in Paris. The agrégation examination is taken in a special field and qualifies successful candidates to higher-paying posts. The CAPES (Certificat d'Aptitude pour l'Enseignment Secondaire), the means of entry for most, is a somewhat less rigorous examination that qualifies one for lower-status posts. Assistant teachers to fill shortages are taken from students preparing for these exams or are hired contractually.

Low salary levels have aggravated recruitment problems and the teacher-student ratio has been dropping. But there is roughly one secondary-school teacher in the urban area per twelve students and one primary-school teacher per twenty-five students, which is a relatively high ratio.

The Politics of Education

Education is one public service in the urban area and throughout France that is highly politically charged. Most private schools are Church schools and the pro-Church versus anticlerical cleavage affects major educational issues. Critical battles between the government and communal authorities in areas loyal to the Church school have arisen over required expenditure on national schools. The prefect has the power, of course, to inscribe such expenditure in the local budget if it is not voted in, but the municipal authorities can harass the public-school administration considerably in exercising maintenance responsibility.

The issue of State aid to private education
has always been a controversial one in the national
political arena. Private schools prepare students
for government examinations and are subject to government inspection. Aid arrangements, however,
have fluctuated with national administrations. At
present, the "Debré Law" of 1960 allows the national government to enter into contracts with private
schools agreeing to assume major teacher salary
costs in return for heightened control over the
teaching program and teacher quality. Almost all
private schools, under financial pressure, have
signed such contracts.

Appraisal of Education
in the Region

Education in France is of exceptionally high
quality, and that provided in the Paris urban area
is above average for France. Present problems in
the region are related to school construction and
expansion of teaching staffs to keep up with growth
of school-age population, particularly intense in
suburban communes, and to meet the needs created by
the current rise in compulsory education age from
fourteen to sixteen. Obsolescence of school plant
and low supply of supplementary facilities, such as
for sports, are deficiencies felt to some extent
throughout the nation. Generally, however, the
French education services have not in the past been
geared to continuing expansion, for the nation had
stable (sometimes declining) population before
World War II. Postwar growth, urbanization, and
now the rise in the compulsory number of years of
attendance have thus posed a considerable challenge,
to which the traditional organization has responded
fairly rapidly.

Neither cost nor classroom capacity keeps children out of school. Over 95 per cent of compulsory
school-age children regularly attend school. Today
almost 70 per cent of the students in the Paris Region continue their education beyond the compulsory
age. Public education is free, and a portion of

primary-school books are provided by the ministry.
Distances traveled to school and students per
classroom, however, are the symptoms of shortages
in specific areas, particularly in Seine-et-Oise.
Average primary-school class size in the City of
Paris is 30. The average in the rest of the
agglomération is 33 to 35, with 40 common in fast-
growing towns. In Seine the average was 30 in 1945,
35 in 1955, and 33 in 1962. Average secondary-
school class size is from 35 to 45 in the region.
Mobile classroom facilities, which are easily assem-
bled and disassembled, have been used throughout the
region as stopgaps during school construction.

Planning and Investment

Education planning has been the responsibility
of the Ministry of Education with over-all investment targets included in the national economic plan.
It remains predominantly the responsibility of the
ministry, but regional investment targets for school
building are included in the 1962-65 national plan
and the Paris district four-year plan.

District targets based on projections of new
dwelling-unit construction were to build 6,350 new
primary-school classrooms in the four-year period
and 11,300 (of which 6,200 are for Seine-et-Oise)
between 1966 and 1970. Secondary-school needs are
put at 145,000 new pupil places in 1962-65 and a
total of 911,400 pupil places between 1964 and 1970.*

The district plan proposes a total capital expenditure in the region during 1962-65 to meet these

---

*The rise in the compulsory school age is at
the roots of this requirement. Attendance until sixteen is to be fully enforced by 1970. Of the 145,000
places proposed, about 45,000 are to replace obsoles-
cent, overcrowded, and temporary facilities and
100,000 are to accommodate age-group growth and increase in the proportion of children attending secondary school.

targets of 2.7 billion francs (1.2 billion for primary schools and 1.5 billion for secondary schools).
The national plan had allocated about 512 million
francs of national funds to primary-school building
in the region (to meet 80 per cent of construction
costs, excluding land and materials) and 970 million francs to meet 85 per cent of total construction costs for secondary schools in the region.
The district plan points out that not only were
these credits probably inadequate to meet the targets, but also commitment of funds by the national
government was behind schedule. Rather than the
necessary 240 million francs per year in State
funds called for by the national plan for secondary schools, program authorizations for the
region were 150 million in 1961 and 100 million in
1962 (about 10 per cent of such authorizations for
the nation).

Plan implementation is contingent on the flow
of State credits. The district did not establish
a works program or intergovernmental financing proposal for schools as it did for water and transportation. The national-government portion of educational finance has been growing and will continue
to do so (about 90 per cent of all school construction is financed by it). Decisions by national
authorities--particularly the Ministries of Education and Finance--are crucial. As the Ministry of
Education budget comprised 17 per cent of the total
national budget in 1965 (compared with 7 per cent
in 1952), financial policy bears an important relationship to the rhythm of school construction.
Financial planning is not integrated into education
planning but enters into consideration at the program authorization stage.

Notes to Chapter 5

1. District de la Région de Paris, Recueil de
documents relatifs au budget de 1964 (Paris: District
de la Région de Paris, 1964), p. 47 (author's
translation).

CHAPTER **6** COMPARATIVE TRENDS AND
PROBLEMS IN URBAN
ADMINISTRATION

PATTERNS OF ADMINISTRATION
IN THE PARIS REGION

The most striking characteristics of the government system of the Paris Region are its complexity and its high degree of centralization. Each
major urban service that we have examined in detail
engages a great many agencies of several levels of
government, linked in a network of controls, supervision, joint participation, and continuing communication and negotiation. Each government level--
national, departmental, and municipal--operates
through traditional administrative subdivisions.
There are ministries, regional offices, departmental offices, and municipal bureaus with specific
duties in the fields of housing, transportation,
education, water supply, and other local public services. The several agencies dealing with any single
function are highly interdependent; there is little
autonomy of action on any level.

Decision-Making

In analyzing the role of various government
institutions in the Paris urban area, the first
conclusion that emerges is that central-government
authorities exercise ultimate power in major decisions for the region--decisions that control the

commitment of capital funds, recruitment of public
personnel, the structure and procedures of all units
of government, the substance of major policies, the
techniques and methods to be followed by local
administrators, and in many cases the details of
project design.

This power is maintained and exercised by three
formal methods. First, powers of decision or review
in specified areas are reserved to central authori-
ties--generally the prefect, the ministries, or the
prime minister. For example, most decisions of the
Seine and Paris councils require approval by central
authorities or by the prefect, acting as their representative. Local borrowing and budgets, creation
of special agencies or special districts, issuing
of construction permits, public-land transactions--
all are examples.

Second, a host of central regulations define in
detail how local authorities may act. Thus, national law and executive regulations may specify the
structure of a municipal or departmental special
authority or districts, the technical standards of
the services that it furnishes, the composition of
its governing board, and the form of contracts into
which it may enter. A large proportion of local-
authority expenditure is closely regulated by virtue
of legal definition of obligatory service standards.

Third, the ministries have great discretionary
power by virtue of national control over financial
resources, particularly capital funds. Every substantial public work in the region requires the
joint financial participation of several levels of
government. National grants for public-works projects ordinarily require decisions from both the
technical ministry concerned and from the Ministry
of Finance; the process requires approval of specific projects, project design, and other matters.

Each proposed public undertaking goes through
a maze that meanders from local units through the
prefect, in and out of a host of operating bureaus

and advisory committees, to one or more central ministries. Thus, authorization of work on the regional transit system required formal decisions by
the Ministry of Finance, the National Economic and
Social Development Fund, the Interministerial Committee for the Paris Region, the Ministry of Public
Works and Transport, the regional transit authority,
the national railroad corporation, and the District
of the Paris Region.

Planning

The District of the Paris Region is now the
major planning authority for the entire urban area
and the focal point for formulation and articulation of plans for regional urban development.
Prior to its establishment, continuous and comprehensive multifunctional planning did not exist.
Even here, the district occupies a middle position.
On one hand, its planning decisions must be fitted
into the framework laid down by the General Planning Commission and other central authorities. On
the other, much of the detailed town planning is
done by, or with the cooperation of, regional and
departmental construction and urbanism services.
The district's freedom to maneuver is therefore
highly circumscribed.

Management

As to the execution of planned projects and
management of local services, we find that the
central government has depended in large part on
the municipalities and the département services.
Increasingly, however, special operating authorities created by local or national authorities are
taking on these activities, substituting functional
specialization for local geographic specialization.
This trend has been noted, for example, in the
studies of urban transportation, water supply, and
public housing. While such transfer of functions
removes them from direct local control, it intensi-
fies national-government operating control through
the elaborate regulations and tight supervision

exercised over special authorities. In the process,
the services are insulated from each other. While
this insulation complicates the problem of coordination, it appears not to have deteriorated the quality
of urban services per se, which is generally high.
The good record can be ascribed to the high level of
personnel qualifications and education and to tradi-
tions of honesty, regularity, and practicality.

PARIS EXPERIENCE IN AN
INTERNATIONAL CONTEXT

This is one of twelve studies of the way in
which urban regions are being organized to cope
with the explosive demand for public services stem-
ming from rapid urbanization.* A major observation
that emerges from comparative analysis is that the
nature of demands, and their fundamental implications for administration, are similar everywhere.
The type and intensity of administrative problems
and response to manifest needs, however, vary
widely among urban areas with organizational, political, and socio-economic factors.

The similarity of demands on government reflects
common characteristics of urban life and form in that
they are directly related to increasing population and
population densities, rapid economic changes, and

---

*The other case studies analyzed to date are
Calcutta, India; Casablanca, Morocco; Davao, Philip-
pines; Karachi, Pakistan; Lagos, Nigeria; Lima, Peru;
Lodz, Poland; Stockholm, Sweden; Toronto, Canada;
Valencia, Venezuela; and Zagreb, Yogoslavia. France
is classified as high (of high, medium, and low) in
gross national product per capita and annual growth
of that aggregate; literacy rate; employment in industry as a percentage of working-age population; and
number of votes in national elections as percentage
of voting-age population. It is in medium range with
respect to population in towns of 20,000 or more.

changing land-use patterns. The most severe urban problems concern housing, transportation, water and sewage facilities, schools and hospitals. Shortages and deficiencies in these areas plague Paris along with other urban areas.

There are other needs that are common to all areas, but are most severe where the contrast between rural and urban life is greatest and urban migration rates are highest--needs for education and training for new job skills and life styles, for new governmental health and welfare services, for intensified law enforcement. Problems of pollution control and land-cost inflation tend to be greater the larger the urban complex.

Because governmental response tends to lag behind demand, unmet needs in all areas accumulate at a faster rate than population. The per capita supply of facilities in housing, transportation, water supply, and other utilities has in fact been declining. This has been true in the Paris Region throughout most of the postwar period. But because Paris had accumulated an extensive network of facilities (transit, water supply, schools, and parks), it still ranks high among cities on any absolute scale of public facilities per capita.*

Demands on government posed by rapid urbanization have direct implications for administration that can be analyzed as to the degree and nature of structural adaptation implied. The first administrative implication of urbanization is the necessity for coping with change. Urban growth comprises social and economic change, which produce not only quantitative increase in demands for government

---

*There is a significant correlation between per capita public-service levels in urban areas and national wealth in the areas studied but no significant relationship that we can ascertain between wealth and the relative increase in public-service levels.

activity but also substantive modification in the
types of activity called for. Administrative responses to urban growth are contingent upon several
types of change within government: quantitative
increase in public investment; policy innovation;
and creation of new--particularly developmental--
activities. Between 1945 and 1959, most of the
energies and expenditures of government for Paris
were devoted to management of existing programs,
with consequent increases in gaps between governmental output and urban problems. In fact, one
objective of national policy has been to restrict
investment in the Paris Region in the interest of
economic decentralization and greater dispersion
of population throughout the country. Still another contributing factor was the separation of the
power to act on urban needs, centered in the national government, from the immediate interests in urban
improvement, centered in the inhabitants of the
Paris Region and their local governments.

Because no authority was responsible for meeting pressing needs of the region, or even for articulating them, and because the political process
did not accomplish these functions, the central government's neglect of growing regional needs received little notice. A communal mayor in Seine has
concluded, "Because of their lack of understanding
of the problems facing the Paris urban area, the
supervising authorities are in large part responsible for the lag in investment." On the other
hand, few local political authorities looked beyond
their immediate jurisdictions toward the region's
large-scale needs or attempted to bring pressure on
the national government to meet them. In short,
there was little pressure for change.

We would, of course, expect that the high degree of both centralization and fragmentation of the
power to act, combined with inadequate means for
expressing growing urban demands, would be associated with low rates of investment and innovation.
The most conspicuous deficiencies in this respect
we have found are in areas with relatively weak

local governments and no regional or metropolitan
institutions (such as Lagos, Lima, Paris prior to
1960), or in areas with metropolitan institutions
that have only weak planning, administrative or
financial power--such as Lodz, Casablanca, Valencia,
and Davao. In the case of Lodz, the existence of
both regional planning and metropolitan-wide governmental organization did not overcome the effect
of highly centralized authority, national policy to
limit growth of the area, and lack of forceful ex-
pression of specific urban needs by political organizations and interest groups.

It appears, further, that identification of
needs, innovative proposals, and pressure for governmental change emanate more frequently from elements within the government and the service bureaucracies than from special-interest groups or political organizations. This has been true of Paris.
The new thrust of activity that is identifying problems and formulating programs for the region is the
result of presidential leadership against a back-
ground of developing concern in some of the ministries and a growing clamor from local-government
agencies, reinforced by the work of the Paris District after its creation in 1961. It is notable
that district officials, far from feeling beset by
community pressures, are attempting to create them
through public-relations efforts directed at arous-
ing public concern for urban needs.

Second, demands on urban government usually
pose the need for more expenditure, both for current
services and capital facilities, and therefore more
funds. But fiscal limitations, particularly on
raising capital funds, curb governmental responses
to urban needs in every area studied. Urban needs
run into competition for funds from other national
interests, such as investment in industries, agriculture, or other national development projects
such as dams and roads. Such competition is the
more direct and serious in that most urban areas
depend heavily on national governments for financial resources; of the twelve urban areas studied,

only Stockholm, Toronto, and Zagreb are exceptions.
Even in these three, local authorities are ultimately subject to national control over the availability
to them of public capital and private credit. National fiscal decisions everywhere are central factors in urban development, even in nations striving
for a larger degree of governmental "decentralization."

Paris illustrates one aspect of the effect on
urban development of national fiscal policies. The
Ministry of Finance has consistently taken a conservative position on development proposals such as
for housing, water supply, and transit. The expressed justifications for this position have been
grounded in macroeconomic and fiscal policies.

Local revenues generally do not expand at a
rate commensurate with growth of population and
urban needs. Local officials in the Paris Region
complain of the inelasticity of local revenues (as
local officials do almost everywhere). While the
leeway for increasing revenues by local-authority
decisions is narrow in Paris, as in most areas,
local authorities have not made full use of existing fiscal powers (a situation also observed in
such cities as Calcutta, Davao, Lima, Stockholm,
Toronto, and Valencia). In the Paris Region, as in
others, political resistance to higher user charges
and increased local taxes has been stronger than
popular demand for urban improvements. This is
particularly the case for programs having as their
main objective the redistribution of income, such
as welfare and locally subsidized housing programs,
which in most cases have been slowest to develop.

In general, it appears that higher-level governments can more readily increase taxes to finance
urban development than can local governments. The
communes in the Paris Region have been loath to
substantially raise rates of the "additional centimes," but the national government was willing to
levy the special tax for the Paris District.

Even so, given the slow response of the national government to local needs, it appears that significant progress in meeting deficiencies in urban facilities and services in the long run must depend not only on national-government support but also on heavier use of local taxes and user charges, and local borrowing. As to borrowing, powers of local governments in the Paris Region historically have been subject to stringent legal restrictions; these have been somewhat relaxed in recent years. In other countries, limitations on borrowing powers are common to most local governments, though the limitations arise in many not so much from legal restrictions as from the limitations of the markets for municipal securities combined with low municipal credit ratings. This is particularly true of cities in the less developed nations and of smaller residential suburbs.*

The imbalance between public-service needs and tax base in various municipalities within an urban area is another common problem of urban public finance. The first approaches to it in Paris consist in the district tax, district grants, and the regional equalization fund.

A third implication of demands on urban government has to do with the specialized managerial, planning, and technical skills required to provide large-scale urban services. Modern transit and water-supply systems, housing construction, social programs, urban development, and renewal projects are all highly complex and technical activities requiring trained staff, advanced engineering technology, and special equipment.** One type of problem this poses is that of fulfilling personnel requirements. The personnel-recruitment problem differs as

---

*In Toronto, municipal-government borrowing is handled through bonds issued and backed by the metropolitan government, which lends its credit to its constituent municipalities.

**Personnel and technical limitations are far

between less developed nations where skills and
technology are generally lacking, and the highly
developed countries where the problem is one of
diverting skills and technology from other sectors
to urban-development purposes.

While France is on the high end of the developmental spectrum, has established relevant education
and training programs, and has a strong civil-
service tradition, municipalities nevertheless have
had serious staffing problems. These have arisen
despite postwar regulations designed to raise qualifications and increase mobility of government personnel serving large cities and towns. Preceding
chapters have noted that technical weaknesses have
arisen in the public-housing industry and in water
and dam services. As a means of enhancing the
attractiveness of top administrative posts in the
City of Paris and the Département of Seine, they have
recently been incorporated into the national civil-
service system.

Another means of supplementing and strengthening local-government personnel is through the technical assistance provided by higher-government levels. Thus, Paris suburban municipalities have come
to depend on départements for many technical operations such as construction of reservoirs and highways. (In the Stockholm area, the central city performs many complex tasks for the suburban governments; and in Zagreb, most major technical responsibilities have been delegated by the communes to
the metropolitan government. In several areas,
national authorities prepare technical plans for
local authorities.) In addition, some technical
services of the city, département, and ministries
have been consolidated in the Seine prefecture.

---

tighter brakes on decentralization of governmental
responsibilities for urban than for rural administration.

Current reorganization, which redraws département boundaries and separates out the city administration, will undo, however, the structural consolidation between the city and département. Another instance of consolidation of central-city and regional agencies is to be found in the Prefecture of Casablanca. In Lodz and Davao, local agencies manage some national services under the directing authority and aid of the ministries, which is another pattern of administrative consolidation. City department heads in Davao are nationally appointed officers.

Two other factors arising from the increasing scale and complexity of urban governmental services frequently pose problems. First, maintaining desired levels of efficiency may require improved procedures and analytical techniques such as program evaluation, data processing for tax collection, central purchasing, and inventory controls. But bureaucratic resistance tends to slow administrative improvements in Paris, as well as in other urban areas, and there has been little research on possibilities for improvement of administrative techniques. Second, the growth in the size and number of urban activities, as well as their increasingly technical nature, increase the proportion of practical decisions that must be made within the bureaucracy, as is the case in Paris. While some local politicians feel that this trend has already gone too far, relationships between political officials and civil servants in France appear generally to be flexible and acceptable to most. However, in many cities where there is close political control over contracts, minor appointments, and technical decisions, the situation is quite the reverse. The tendency to ignore the advice of a specialized staff is notable in several of our study areas, particularly where strong councils predominate and where relationships between administrative and political roles have not been established by tradition. The resultant problems are most severe where political competition is focused primarily on appointments and contract awards rather than on directions of

policies and programs. In general, the more restricted the discretion of the bureaucracy in day-
to-day decisions, the less the energy of political
officials or bureaucrats devoted to policy matters.

   At the same time, the greater the magnitude of
bureaucratic decision-making, the greater the resources required by political officials in order to
maintain control over important policy issues. In
some areas with strong bureaucracies, including Paris,
there are certain mechanisms that may be used by
political officials to enhance their capability to
exercise general control over the bureaus and services for purposes of policy direction and coordination. Cases in point are professionally trained
chief administrative officers (the secretary-general
in France) and top-flight staff (the staff cabinets in France) serving political officials. There
appear to be limitations, however, on the degree to
which these serve the officials to whom they are
responsible, because there is always the possibility
that administrative officers will form stronger alliances with the bureaucracy than with their legal
superiors. In response to this problem, the secretary-general of the city administration in Zagreb
has been made subject to periodic reappointment by
the city assembly. Another approach is to establish a second layer of politically responsible
officials overseeing administration. The assistant mayors in a French urban commune and political
commissioners in Stockholm (who have greater formal powers over department heads) are cases in
point. Specialized executive boards supervising
one or a few departments are other examples found
in Stockholm and Zagreb.

   In general, it appears that policy control and
coordination of a strong bureaucracy are far more
difficult for an elected legislative body to exercise directly, than for a functionally specific,
strong executive--albeit an individual or board--
independently selected or chosen by the council.
The executive officials in the Paris Region--mayors,
and particularly prefects--are relatively strong by

comparison to those in other systems. This pattern
raises issues, however, as to the prerogatives of
elected councils in the local-government system.

A fourth aspect of urban problems with administrative implications is time dimension. Most major
capital projects in urban areas have an effective
life and span of influence on related urban development of twenty years or more, whereas they are frequently designed to meet only immediately foresee-
able needs and their impact on development is largely neglected. The perspectives of government are
often limited to one or a few fiscal years. In con-
sequence, facilities are often obsolete before completed, do not produce service levels contemplated,
and have unforeseen side effects. In periods of low
capital investment, as in Paris during the postwar
period, these problems are not apparent. Such inadequacies are not identified until actual service
needs are projected by planning studies, or, in ret-
rospect, after major projects are completed and
evaluated. Adjustment to the time-dimension aspect is seen in the integrated planning efforts by
the District of the Paris Region, which focus on
four-, twelve-, and twenty-year periods, and the
institution of four-year capital budgets for the
City of Paris and Département of Seine. The crucial aspect of this adjustment, however, is plan
implementation. Examples of long-range plans with
little or no impact on budgetary decisions or program design are plentiful. Hence, evaluation of
the new policy dimensions in Paris must wait some
years.

The success or failure of the effort to integrate plans of different time dimensions will be
of particular interest. In general, urban plans
dealing with land use and population can cope with
longer time periods than those dealing with projects and investments, if for no other reason than
presently limited techniques of economic and policy
forecasting. The effort to formulate four-year
project plans implementive of twelve-year development plans and twenty-year land-use plans is a

unique attempt to integrate various types of planning.

In some areas, plans themselves have been based on inaccurate forecasts and analyses in which case they do little to mitigate the problems cited. The development of historical statistical data and techniques of policy research is essential to rational long-range planning and policy formulation. Paris is comparatively well equipped in this respect, particularly by virtue of the several public and semipublic institutions that have been created to collect and analyze statistics for the region and to undertake planning research.

Finally, and by far the most difficult aspect to cope with, the nature of urban life and form poses problems relating to the organization of general urban government. The first dimension of organization for which urban growth has direct implications is geographic. As scholars and practitioners alike have long recognized, urban growth produces a socio-economic unit that generally bears no territorial relationship to traditional governmental jurisdictions and political arenas. As has been manifest in Paris, many urban problems are metropolitan in scale. Some of the newer activities of government in the Paris Region are by nature metropolitan--such as regional transit and land-use development planning. Others are local but impinge upon common regional resources--such as water supply. Thus, the use of spring water throughout the region by the City of Paris system circumscribes the resources available to recently urbanized portions of the region. Still other local activities affect the demand for services in other parts of the metropolis. Moreover, the uneven distribution of public facilities among communes of the region is increasingly a cause of local complaint.

As a result of these factors, intensive metropolitan relationships are found in all the areas studied, ranging from continuous intermunicipal conflict and negotiation to established metropolitan

government. A trend from that first end of the
spectrum to the other, or institutionalization of
metropolitan relationships, is common. In some
nations, such as Yugoslavia, Poland, and Morocco,
traditional units of subnational administration
(districts or provinces) have been adjusted to encompass metropolitan complexes and have been endowed with special responsibilities with respect to
urban services. For others, such as Toronto,
unique tiers of metropolitan government have been
created. Davao and Valencia are endowed with traditionally large municipal units, which will encompass urban growth for some years to come and are
therefore in effect metropolitan cities. In the
remaining areas, including Paris, metropolitan
planning units and special public-service authorities have been instituted.

     In addition, in Paris special mechanisms for
administrative coordination in the region have
been instituted. The option of creating a metropolitan département has not been chosen, but intensified needs for regional coordination arising from
the increase in the number of départements within
the region have been recognized in establishing the
regional prefect. In addition, limited regionalization of public finance and certain plan projects are
aimed at reducing public-service imbalances among
communes. While this approach may overcome problems of geographic scale in urban organization,
evidence from other areas indicates that it may
intensify problems of functional organization.

     This brings us to a second dimension of the
organization, for which there are important urban
ramifications. Increased specialization and interdependence in the urban economy extend to the economic functions of government, as we have noted. The
quantitative, qualitative, and spatial relationships
among water supply, sewage, housing, transit, education, employment, and other urban functions that are
substantially supplied or influenced by separate
governmental bureaus and hierarchies are well docu-
mented. Specialization without concomitant

mechanisms of interdependence (or, as it is commonly referred to in public administration, "coordination") produces disjointed activity that has been described as fragmentation. The number of specialized bureaus, services, and authorities multiplies with urbanization. The contrast between an administrative structure consisting of mayor, schoolteacher, and policeman in the rural communes of the Paris Region and the huge and complex structure for the City of Paris and Département of Seine illustrates in space this trend in time under urbanization pressures. Methods for effectively correlating the activities of the manifold public agencies involved in not only management of public services but also various development efforts are the central concern of current governmental reforms in Paris.

In highly centralized systems of government such as that for Paris, a major approach toward interservice harmony is formulation of comprehensive policies for the region by the national government and appropriate coordination of the policies of national agencies as they act with respect to the region.

The apexes of decision-making for the Paris Region are in various branches of the national ministries over which there was traditionally little policy coordination or strong executive control. In the past, the response by the national government to the complexity of administrative machinery in the Paris Region has consisted mainly in intensification of central controls with neither coordination of the controlling hierarchies themselves nor shared policies among them.

The effects of this functional fragmentation of decision-making were not of tremendous import during the years of low public investment in the region, but nonetheless, they did limit the efficiency of urban facilities by comparison with present goals. The transit system stopped not far outside the city without connection to railroad services beyond. New housing developments rose in

areas with already overcrowded schools, while classrooms in other communes were underused. Certain
sewage and disposal practices reduced the potential
of certain sources of potable water. The most serious instances of disjointed administration were
mitigated by compensatory action; thus, for example,
education authorities concentrated school construction subsidies where overcrowding had developed,
and water agencies turned to alternative sources or
treatment methods.

The case studies indicate, however, that compensatory and incremental decision-making does not
suffice at times when large investment and development efforts are undertaken with the purpose of
achieving a great leap forward in the efficiency
and convenience of urban life. Simultaneous investment decisions cannot, by virtue of scheduling, be
incremental. Moreover, when society or government
adopts the goal of consciously choosing future
patterns for urban development, approximation of
these patterns is contingent upon comprehensive
policies and coordinated decision-making. Therefore, while the District of Paris project plans and
regional land-use plan provide a new policy basis
for potential coordination of decision-making on
the one hand, their goals actually create the need
for such coordination, on the other. Creation of
the regional transit system, establishment of new
town centers, and scheduling construction of community facilities to population growth and distribu-
tion--all goals of the plans--will require new
styles of decision-making cutting across traditional agency boundaries. The translation of comprehensive concepts into complementary development depends ultimately on the effective and continuous
exercise of pressure to this end through both the
regional prefect and the Interministerial Committee
for the Paris Region.

The case studies suggest the general conclusion that coordination and shared policies are in-
frequently achieved when substantially independent
bureaucratic structures remain unchanged, except

where strong and concentrated political control is
exercised or for short periods of special leadership. The increase in the number of départements
and regional special authorities (if the latter are
created) in the Paris Region will add to the number
of institutional actors to be correlated. The regional prefect faces an even more difficult task
than a normal departmental prefect; and the latter's
ability to influence service chiefs has been subject to severe extralegal constraints, as we have
seen.

The effects of interservice relationships on
program achievement appear common to large urban
complexes where government provides substantial public goods and services. The specific organizational
responses to these relationships, however, vary with
the governmental structure, especially with the distribution of essential decision-making powers with
respect to public-service expansion and development
projects.*

In Casablanca, Davao, Karachi, Lima, Lodz, and
Valencia, where national authorities play major
decision-making roles, interservice coordination
depends, as it does in Paris, upon mechanisms linking persistently independent national hierarchies
at the metropolitan or regional level. As we have
demonstrated with respect to Paris, the arena for
coordination is metropolitan, but the institutional
components to be coordinated are in large part
nationally controlled. On the other hand, in areas
in which there are general metropolitan governments
within which are situated major powers over urban

---

*While in all cases higher-government authori-
ties control the amount of investment capital available for urban development, the essential variable
here involves who controls the nature of the projects to be undertaken and their location in the
urban area.

services and development, the degree of interservice coordination depends, first, upon the development within that government of comprehensive policies and executive influence over operating agencies; and, second, upon the coordinating powers of
that metropolitan government over lower-tier municipal units, if these exist. Thus, the active organizational issues in Toronto and Zagreb today concern metropolitan planning, the intensity of control that metropolitan-government units should
exercise over special authorities, methods for
strengthening general legislative and executive
control over the metropolitan bureaucracy, and the
allocation of powers between metropolitan and municipal units. Although Lodz, Valencia, Casablanca,
and Davao have metropolitan units, because decision-
making powers vis-à-vis urban-development functions
are concentrated in the national bureaucracy, their
organizational problems in this respect resemble
those of Paris more than those of Zagreb and Toronto. In all of these four, however, the metropolitan institutions are attempting to interrelate
national agencies operating within their territory.

Finally, where there is no metropolitan government, and crucial powers over various urban functions vest in municipal units, interservice coordination is inextricably linked to municipal relationships. Proposed metropolitan organization for
Stockholm contemplates establishment of a metropolitan county that would have operating responsibilities in transportation and health care, but would
have mainly planning and coordinating powers over
municipalities in other service categories. Its
purposes would be both geographic unity and functional coordination.

These prototypes have been simplified for
analysis, of course. Municipalities and départements in the Paris Region do play a substantial
role in urban administration, and the purposes of
regional organization encompass systematizing their
relationships, particularly in the district council
and through district plans, as well as systematizing

the relationships of national-service hierarchies. To date, however, the emphasis has been on the latter.

Planning is an important component of systematizing these relationships in all of the variant structural patterns. The coherence of urban-development goals and government approaches to urban problems is associated with the incidence of research into the economic and social relationships of programs, on the one hand, and formulation of comprehensive policies that can (but may not) serve as reference points for complementary decisions, on the other.

There is a second stage of interservice relationships that impinge upon program achievement. We have been discussing cohesiveness and coordination of basic development policies and investment decisions. The stages of day-to-day-management and work scheduling entail another set of relationships, primarily bureaucratic, that vary with cultural and personality factors, as much as with political and organizational factors. Paris has a comparatively good record of bureaucratic cooperation in this respect, in which the achievement orientation of public personnel and education and recruitment procedures play no small part. In addition, there are several formal mechanisms in French administration that intensify interagency communication on practical problems. These include special committees and work groups, circulation of work proposals, and mobility of members of the senior civil-service cadres. Hence, Paris is relatively free from some of the most absurd results of functional fragmentation found in many other areas—particularly those in less developed nations—such as completed housing units that remain empty for want of water connections or transportation access, delay of repairs by one agency of the equipment of another, and operational stalemates on programs involving several agencies.

Intergovernmental Relations

The discussion thus far has shown that one of
the crucial variables of government structure that
underlies the nature of response to the various
administrative implications of urbanization--innova-
tion, expansion of public finance, coping with technical complexity, broadening of time perspectives,
and adjustment of organizational relationships--is
the degree of centralization (or, conversely, decentralization) of powers in urban government. National and/or state governments are important participants in developmental aspects of urban administration in all of the urban areas analyzed except
Zagreb, and in most areas their relative importance
is increasing in spite of the near universal existence of expressed "decentralization" policies. In
Zagreb, centralized political organizations, integrated national-local planning, national allocation
of aggregate financial resources available to local
authorities, and extensive legal regulation provide
a framework of control within which metropolitan and
communal authorities have exclusive decision-making
authority over urban projects. In the other areas,
veto powers with respect to major projects by virtue of control over capital finance are the minimal
common role of higher government.

The pattern of intergovernmental relationships
found in Paris is common: Several institutions at
two or more levels of government are involved in
each public-service category, and each must take
specified action to move a project from conception
to execution.

As has been demonstrated in study of Paris,
the opportunities for stalemate within so plural
and complex a decision-making structure are plentiful. Instances of the time-consuming and sometimes
paralyzing involvement of a multiplicity of agencies, committees, and central authorities in the
details of each public activity have been noted with
respect to land control, water supply, transportation, and school construction.

To accommodate this pattern, intergovernmental
relationships tend to expand from strictly formal
modes of control to more intensive and flexible
communication. An efficient partnership approach to
urban development and public-service provision depends upon negotiation, bargaining, and exchange of
information, on the one hand, and leadership and
clear lines of authority for resolving disagreement
and getting things done, on the other.

Recent innovations in Paris have intensified
intergovernmental communications. The planning
efforts of the district have essentially been communications processes engaging local representatives,
departmental and national bureaucrats, and the national political leaders in formulation of some basic
consensus on policies and programs for the region.
Above all, the delegate general of the Paris District and his staff have actively promoted the plan
findings on a continuing basis and formed a link
between local officials and the Interministerial
Committee. As regional prefect this officer may
have greater legal base from which to exercise his
persuasive powers to get things done in the region.

On the other hand, none of the modifications
in government for the region have struck at the
heart of administrative complexity by consolidating
authority. Aimed at formulating and implementing
comprehensive urban-development policies, on the
one hand, and increasing operating efficiency, on
the other (by reducing the size of départements and
upgrading personnel), the reforms leave fundamentally
untouched the unclear and overlapping lines of authority that plague decision-making processes of
middle range.

In Paris, as well as in other centralized systems of urban government where higher authorities
not only maintain veto powers over major actions
but also are involved in a myriad of decisions of
lesser import, it is extremely difficult to identify
key officials or agencies who can, in effect, re-
solve conflict and get routine matters taken care

of with reasonable dispatch.

Complicated and centralized allocation of responsibilities and resources is compounded by detailed formal controls over the exercise of local prerogatives. The administrative controls exercised in the Paris Region undoubtedly are greater than necessary to protect national interests, promote national policies, and assure legality, given the appointment of city and département executives and the circumscribed powers of local authorities.

It appears from the case studies generally that detailed and overlapping checks on local-authority actions do not add up to effective net control. In many of the developing areas studied, illegal or irregular action on the part of local officials is not diminished by proliferating supervisory red tape. Supervising authorities are often overburdened with routine approvals and have neither the time nor the inclination to attack the roots of corruption or mismanagement until crisis periods are reached. Crisis points are frequently cyclical. They result in suspension of local government, token reorganization and additional detailed controls, re-emergence of old problems, and new crises.

The incidence of diversion of public benefits to private purposes is low in Paris, as well as in Stockholm, Toronto, and Zagreb. Comparative analysis suggests that the conditions for this fortunate characteristic are rooted first and foremost in culture, the nature of political conflict, and the characteristics of personnel education and recruitment. The evidence suggests, however, that some modes of control are more efficacious than others and that elimination of the extraneous controls contributes to administrative efficiency.

First, strong and objective controls over fiscal transactions exercised to assure legality are among the most important and are highly effective in Paris, where officials of the Ministry of Finance who are endowed by tradition and training with

high integrity and status are posted as treasurers
within local governments, national ministries, and
many public authorities. A parallel is found in
Yugoslavia, where the fiscal transactions of public
bodies, including local authorities, are carried
out through branches of the national bank, where
they are subject to continuing check by its social
accounting service.

Second, the general supervising official over
local government in France is a member of the prefectoral corps who is given relatively stable
assignments, is generally respected for his integrity, and is not subject to arbitrary political
dismissal. By contrast, there have been three prefects in Casablanca in three years; and in many
other areas, local leaders have recourse through
political channels to obtain the removal of inspectors or supervisors who meddle in controversial
affairs. Third, controls over local authorities in
France and Yugoslavia are in large part embodied in
legal provisions. To take an example, when a local
council creates a public corporation or executes a
contract in Paris, it must do so in accord with detailed legal specifications. The discretionary
latitude of supervision over these acts is thereby
reduced, prior approvals and clearances can be dis-
pensed with, and stable criteria are established for
local authorities to follow. By contrast, in some
areas each and every contract and charter is subject
to discretionary review by central authorities,
with consequent slowing down of administrative processes and uncertainties on the part of local officials as to proper procedures.

THE FUTURE PROSPECTS
FOR PARIS GOVERNMENT

The establishment of comprehensive plans for
the development of the Paris Region and its public
facilities and the attempt to translate the plan
goals into reality will undoubtedly bring to light
new issues and lessons. To date, the present

government of France has concluded that simplification and coordination of administrative structures
is a crucial means to achieving its goals of planned
and balanced urban development. It has also decided generally that metropolitan areas should be
given jurisdictional status for these purposes.
In the fall of 1966, the de Gaulle government intro-
duced a bill that would establish metropolitan councils composed of delegates from municipal councils
in four other urban complexes in France. These
councils would not only undertake urban planning
but would execute development projects and manage
fire fighting, public housing, transit, and sanitation services, as well as public utilities and local
functions with respect to secondary schools.

    With respect to Paris, however, the government
is faced with a dilemma in that significant simplification and authoritative coordination of administration implies consolidated power. The political
leaders of France have long feared any consolidation of power in the Paris Region. If the coordinating mechanisms being superimposed over the highly
independent administrative units in the region do
not prove adequate to the tasks of plan implementation, there may be intensified pressures in the
future to overcome these fears, at least to the
point of institutionalizing regional administration.
Former Prime Minister Debré summed up the dilemma
in more general terms speaking in Parliament:
"Theoretically, it is tempting to create above the
département a regional government level, but in a
democracy to create a regional government is to
create at the same time an elected assembly. Each
one of us asks himself with grave concern what the
impact of regionally elected assemblies would be on
national unity."[1]

    These are the pressures that have militated
against organization of general-purpose regional
institutions, in spite of the criticisms that the
Paris District as presently constituted is not
representative and of the recognized problems of
administrative fragmentation. Local authorities

fear, with some basis, that present trends in regional organization will simply add another layer of
hierarchical supervision and more actors in an already crowded administrative arena. If the piece-
meal delegation of powers from central ministries
to their regional field officers continues, and if
independent regional service authorities are created,
administrative powers will migrate from local and
central levels to the regional level without substantial consolidation and simplification of authority.

Many of the local critics of national-government reorganization policies, however, propose not
general regional government, but decentralization
of powers to local authorities. For the most part,
local officials accept the principle that national
authorities should maintain general policy controls,
offer capital loans and grants, and institute metropolitan coordination. But they urge that local authorities be empowered to act within the framework of
agreed-upon policies without detailed review and
approval of every stage of action from budgets to
construction permits and that they be given expanded
fiscal resources to do so.

While some decentralization measures have been
effected in recent years, consisting for the most
part of increases in the powers of the Paris and
Seine councils, they are outweighed by other centralizing tendencies. National expenditure in the
region has been expanding faster than local expenditure. Independent regional authorities include
central appointees, and the regional prefect is, in
the last analysis, the government's man. The effect
of restructuring the départements is difficult to
foresee, for while this provides for smaller administrative and supervising units, it will probably
reduce the influence of the département councils.
All of the centralizing tendencies in the Paris
Region entail not only greater powers at higher
government levels but also greater powers within
the bureaucracy vis-à-vis elected officials. The

already restricted role of locally elected officials
is further limited by regional institutions dominated
by career officials.

The officers on the district staff who are concerned with administrative reform do not see these
issues in terms of centralization or decentralization of powers, but rather in terms of seeking an
expeditious allocation of functions among all levels
to meet the exigencies of urban service problems.
They foresee modifications in assignment of powers
as practical lessons emerge from current experience.
An officer in the Paris District pointed out:

> Governmental structure for a large urban
> complex should not be on a single level.
> There are aspects of urban administration
> suited for municipal and departmental
> scales of activity as well as for regional
> organizations. A good organizational struc-
> ture would provide these various levels
> with a unified policy base and coordination
> of general administration.

In any case, it is clear that national-government reorganization measures to date have been aimed
neither at organization of regional government
nor at decentralization, but rather at informal
coordination of national and local agencies operating in the region. Evidence from other case studies supports the conclusion reached in Paris that
central-government control does not produce coordinated activity in a complex urban area unless
there exists a unity of concepts provided by comprehensive planning and activated by effective
leadership. In other words, coherent policy is a
major prerequisite for high-level coordination, and
leadership is a major prerequisite for implementation of such policy. The current development of a
regional land-use plan and of four- and twelve-year
development plans for the region can form the basis
for a harmonized approach to the region by various
authorities. The pressures on the ministries and
local agencies to gauge their future programs

for the region in terms of these policies are now
being exerted by the regional prefect, the Office
for Town and Country Planning and Regional Action
in the prime minister's office, and in the Interministerial Committee for the Paris Region. Evaluation of the effect of these arrangements must wait
some years, for they are in their infancy. The test
will be whether, in the future, major programs for
housing, transit, and other urban services comple-
ment one another in such a way as to produce the
patterns of urban expansion in the Paris Region set
forth in the major regional plans and whether administrative activities are scheduled in such a way
as to execute these programs smoothly and swiftly.
At this point, one can only question whether the
coordinating powers established are commensurate
with the powers of existing hierarchies and their
persistent tendencies to independence.

In the final analysis, what is most notable
about urban government in Paris is that authorities
have recognized,first, that the urban complex is a
dynamic phenomenon that must be approached from a
comprehensive stance if it is to be guided along
lines that will increase the satisfaction of its
inhabitants; and, second, that efficient provision
of urban services requires diverse organizational
arrangements utilizing local, area-wide, and national structures alike. They are groping for the proper mix of structures, and the results of present
experiments over the next decade should be of essential interest to other urban areas throughout the
world.

Finally, the period since 1945 has been one of
immense complexity and rapid change in France.
While the larger national and international issues
in which French Government and society have been
embroiled are beyond the scope of this work, they
sketch the perimeters of concern with urban-development problems. The exigencies of domestic investment to meet the demands of competition in the
Common Market and of rapid economic recovery and
expansion, together with high commitments in over-

seas areas, have downgraded the order of priority
assigned to improving the urban environment. The
primary concern of the Parisian has been with improvement to his private life after the trying years
of depression, war, and recovery and with the broad
national issues of the postwar period. Preoccupa-
tion with these matters has tended to limit the
citizen's effective articulation of demand for governmental action to meet growing urban-service requirements, and at the same time the national government has not been fully responsive in such fields
as housing, transportation, water supply, education,
and health facilities. Higher standards of living
and the establishment of economic and political stability in the 1960's may help to stimulate local
interest and political participation in regional
issues that would sustain over the next decade the
present efforts to plan and to effect improvement
in urban form and facilities.

Notes to Chapter 6

1.   Journal Officiel, Debats, July 11, 1961,
p. 179.

SELECTED
BIBLIOGRAPHY

SELECTED
BIBLIOGRAPHY

GENERAL GOVERNMENT STRUCTURE AND
ADMINISTRATIVE PROCEDURE IN FRANCE

Ardan, Gabriel. Techniques de l'Etat. Paris:
    Presses Universitaires, 1953.(Deals with prob-
    lems and efficiency in public administration.)

Aron, Raymond. France, Steadfast and Changing: The
    Fourth to the Fifth Republic. Cambridge: Har-
    vard University Press, 1960.

Blondel, J. "Local Government and the Local Offi-
    ces of Ministries in a French Département," in
    Public Administration. London: Vol. 37,
    Spring, 1959, pp. 65-74.

Bourjol, Maurice. Les Districts urbains. Paris:
    Editions Berger-Levrault, 1963. (Description of the
    structure and utilization of special urban dis-
    tricts and controlling administrative regulations.)

Caplat, Guy. L'Administration de l'education
    nationale et la réforme administrative. Paris:
    Editions Berger-Levrault, 1960.

Chapman, B. Introduction to French Local Govern-
    ment. London: Allen & Unwin, 1953.

_____. The Prefects and Provincial France.
    London: Allen & Unwin, 1955.

Crozier, Michel. The Bureaucratic Phenomenon.
    Chicago: University of Chicago Press, 1963.
    (Analysis of behavior in large-scale organiza-
    tions as related to cultural setting in France.
    Includes two case studies, broad discussion of
    government administrative systems, and brief
    comparative chapter dealing with the United
    States and the U.S.S.R.)

Delion, A. G. Le Statut des enterprises publiques.
    Paris: Editions Berger-Levrault, 1963.

Diamant, Alfred. "The French Administrative System,"
    in William Siffin, Toward the Comparative Study
    of Public Administration. Bloomington: Univer-
    sity of Indiana, Department of Government, 1957.

Duverger, Maurice. The French Political System.
    Chicago: University of Chicago Press, 1958.
    (Study of the dynamics of French politics--
    parties, elections, pressure groups, and rela-
    tionships to government and administration.)

Ehrmann, Henry W. "French Bureaucracy and Organ-
    ized Interests," Administration Science Quarter-
    ly. Ithaca: Vol. 5, March, 1961, pp. 534-555.

Escudier, A.-J. Le Conseil Général. Paris: Editions
    Berger-Levrault, 1964. (Practical guide to the
    structure and powers of departmental council.)

Fabre, F.-J., Morin, R., and Serieyx, A. Les
    Sociétés locales d'économie mixte et leur controle.
    Paris: Editions Berger-Levrault, 1964. (Study
    of special public corporations utilized increas-
    ingly by groups of local governments to provide
    urban services.)

Godfrey, E. Drexel, Jr. The Government of France.
    New York: Thomas Y. Crowell Co., 1963.

Goguel, François, and Grosser, Alfred. La Politi-
    que en France. Paris: Librarie Armand Colin,
    1964. (General survey of politics in France,
    parties, pressure groups, and political trends.)

Grégoire, Roger. The French Civil Service.
    Brussels: International Institute of Adminis-
    trative Sciences, 1964.

Langrod, Georges. Some Current Problems of Admin-
    istration in France Today. San Juan: Univer-
    sity of Puerto Rico, School of Public Adminis-
    tration, 1961.

Macridis, Roy C., and Brown, Bernard E. The de
    Gaulle Republic. Homewood, Illinois: Dorsey
    Press, 1960, and 1963 supplement.

Ministère des Finances et des Affaires Economiques.
    Statistiques des comptes des départements, des
    communes et des certains établissements publics
    locaux. Paris: 1960-1964 (annual).

Ridley, F., and Blondel, J. Public Administration
    in France. London: Routledge and Kegan Paul,
    1964.

Soudet, P. L'Administration vue par les siens...et
    par d'autres. Paris: Editions Berger-Levrault,
    1960.

            URBAN AND REGIONAL PLANNING IN FRANCE

Auzelle, Robert. L'Aménagement des agglomérations
    urbaines. Paris: Presses Universitaires, 1961.

Bordessoule, A., and Guillemain, P. Les Collecti-
    vités locales et les problèmes de l'urbanisme
    et du logement. Paris: Sirey, 1956.

Carillon, Robert. Le Financement du developpement
    régional. Paris: Centre national des économies
    régionales, 1964.

Carrière, Françoise, and Pinchernel, Philippe.
    Le Fait urbain en France. Paris: Librarie
    Armand Colin, 1963. (Statistical analysis of
    urbanization and the economic and demographic
    structures of urban areas in France.)

Centre de Recherches Economiques et Sociales. Le
    Développement des collectivités urbaines dans
    l'aménagement d'une région. Paris, 1960.

Direction de l'Aménagement Foncier et de l'Urbanisme.
    Bibliographie: Aménagement foncier et urbanisme.
    Paris: Ministère de la Construction, 1963.

Faucheux, J. La Décentralisation industrielle.
    Paris: Editions Berger-Levrault, 1960.

Godchot, J.-E. Les Sociétés d'économie mixte et
    l'aménagement du territoire. Paris: Editions
    Berger-Levrault, 1960.

Gravier, J. F. L'Aménagement du territoire et
    l'avenir des régions françaises. Paris: Flam-
    marion, 1964. (Study of the changing use of
    space and patterns of urban development in
    France.)

Hackett, John, and Hackett, A. M. Economic Planning
    in France. Cambridge: Harvard University Press,
    1963.

Institut d'Etudes Politiques de l'Université de
    Grenoble. Administration traditionnele et
    planification régionale. Paris: Librairie
    Armand Colin, 1964. (Papers dealing with new
    arrangements for regional planning and changes
    in social administration throughout France.)

Lajugie, Joseph (ed.). Développement économique
    régional et aménagement du territoire. Paris:
    Sirey, 1964. (A volume of essays on spatial
    regional planning in general and as practiced
    in several countries--France, Belgium, Spain,
    the United States, Britain, Greece, and the
    U.S.S.R.

Lamour, Phillippe. L'Aménagement du territoire.
    Paris: l'Epargne, 1964. (Brief description of
    the principles and methods of spatial planning,
    land-use development, and control in France.)

Lavedan, Pierre. Les Villes françaises. Paris:
    Fréal et Cie, 1960.

Poissonnier, A. La Rénovation urbaine. Paris:
    Editions Berger-Levrault, 1965. (A description
    of the laws, regulations, and administrative
    procedures for urban renewal in France.)

Rossillion, P. Les Plans d'Urbanisme. Paris:
    Editions Berger-Levrault, 1960. (A description
    of the laws, regulations, and administrative pro-
    cedures of urban and town planning in France.)

"Urbanisme," Revue Française. Paris: Vol. 32,
    No. 80, 1963. (Special issue on new organiza-
    tions and procedures for urban and regional
    planning in France.)

Viau, Pierre (ed.). Démocratie, planification,
    aménagement. Paris: Economie et Humanisme,
    1964. (A theoretical discussion of planning and
    citizen participation in government.)

                    THE PARIS REGION

Bastie, Jean. Paris en l'an 2000. Paris: Sedimo,
    1964. (A forecast of the growth patterns of the
    Paris urban area, description of major issues of
    urban reform, and outline of capital facilities
    needed over forty years. Includes sketch maps,
    organization chart, and bibliography.)

Belleville, Germaine. Morphologie de la population
    active à Paris. Paris: Librairie Armand Colin,
    1962.

Boutet de Monvel, Noël. Les Demains de Paris.
    Paris: Denoel, 1964. (Examination of infrastruc-
    ture needs, critique of current plans, and pro-
    jections of the growth of Paris Region.)

Chambre de Commerce. Les Services communaux et
    leur coût dans l'agglomération parisienne. 1961,
    multigr. (27 Ave. Friedland, Paris 8.)

Délégation Générale au District de la Région de
    Paris. Avant Projet de programme duodecennal
    pour la région de Paris. Paris: 1963.
    (Draft twelve-year development plan.)

_____. Programme quadriennal pour la région de

Paris. Paris, 1964. (Four-year plan of capital
investments.)

_____. Schéma directeur de la région de Paris.
Paris, 1965. (Regional land-use plan.)

District de la Région de Paris. Bulletin Officiel Du.
    (Trimestriel.)

_____. Recueil de documents relatifs au budget
de 1964. Paris, 1964.

George, Pierre, et al. Etudes sur la Banlieue de
    Paris. Paris: Librairie Armand Colin, 1950.

_____, and Randet, Pierre. La Région parisienne.
    Paris: Presses Universitaires de France, 1959.
    (Descriptive work covering generally economic
    activities, population, infrastructure, and
    national resources.)

Griotteray, Alain. L'Etat contre Paris. Paris:
    Hachette, 1962. (Expounds the thesis that over-
    centralization of fiscal and decision-making
    powers is a major obstacle to solution of urban
    problems in Paris.)

_____. "Un programme pour Paris," Journal de
    Communes (Edition Spéciale). Paris, 1965.

Imprimerie Municipale. Paris, 1960, Paris, 1961.

INSEE. Direction régionale de Paris. Annuaire
    statistique abrégé de la région parisienne.
    Paris, 1961. (Statistical abstract of the
    Paris Region.)

_____. Delimitation de l'agglomération parisienne.
    Paris, 1959. (Report of a study to define the
    urban area; four zones are defined and method-
    ology described. An English translation by A. M.
    Walsh available from Institute of Public Admin-
    istration, New York.)

L'Institut d'Amenagement et d'Urbanisme de la
    Région Parisienne. <u>Cahiers.</u> Paris, 1963-66,
    Vols. 1-5. (Research reports on the Paris Re-
    gion: land use, economic and demographic trends,
    planning documents.)

Léautey, M., and Tafforeau, M. <u>Le Régime adminis-
    tratif du Département de la Seine et de la Ville
    de Paris.</u> Paris: Ecole Nationale d'Administra-
    tion Municipale, 1962.

Legaret, Jean. <u>Le District de Paris.</u> Paris:
    Imprimerie Municipale, 1963.

_____,et al. <u>Le Statut de Paris.</u> Paris:
    Librarie Générale de Droit et de Jurisprudence,
    1958.

Ministère de la Construction. <u>Plan d'aménagement
    et d'organisation générale de la région parisi-
    enne.</u> Paris, 1960.

Pilliet, G. <u>L'Avenir de Paris.</u> Paris: Hachette,
    1961.

Piquard, M., and Viot, P. (eds.). <u>L'Organisation
    administrative de l'agglomération parisienne:
    Le regime administratif des collectivités publi-
    ques de la région de Paris.</u> Nine vols. (Mimeo.)
    Paris: Institut d'Etudes Politiques de Paris,
    1964.

Pourcher, Guy. <u>Le Peuplement de Paris.</u> Paris:
    Presses Universitaires de France, 1964. (Report
    of a social survey of the Paris urban area, par-
    ticularly of migration and mobility and the
    attitudes and living conditions of in-migrants.)

Prefecture de la Seine. <u>La Conjoncture économique
    dans le département</u> (1er Trimestre, 1964). Paris:
    Imprimerie Municipale, 1964.

Régie Autonome des Transports Parisiens. <u>Rapport à
    M. le Ministre des Travaux Publics et des Trans-
    ports, Exercice 1963.</u> Paris, 1964.

ABOUT THE AUTHOR

Annmarie Hauck Walsh is a member of the research staff of the Institute of Public Administration, New York, and director of its International Urban Studies Project. She has participated in a wide range of studies at the Institute dealing with urban affairs, including urban transportation and outdoor recreation resources in American metropolitan areas.

During three years of research in comparative urban administration, she has traveled in Europe and Africa and participated in several field studies.

Mrs. Walsh was formerly editor of Metropolitan Area Problems: News and Digest. She has studied comparative politics and government at Smith College, the University of Geneva, and Columbia University.